Math Essentials

Speed Wheel Drills
for Multiplication

Richard W. Fisher

www.mathessentials.net

Speed Wheel Drills for Multiplication
Copyright © 2021, Richard W. Fisher.
Printed and bound in the United States of America.

All rights reserved. No part of this book may be reproduced or transmitted in any form or by any means, electronic or mechanical, including photocopying, recording, or by an information storage and retrieval system—except by a reviewer who may quote brief passages in a review to be printed in a magazine, newspaper, or on the Web—without permission in writing from the publisher.

For information, please contact Math Essentials, 469 Logan Way, Marina, CA 93933.

Although the author and publisher have made every effort to ensure the accuracy and completeness of information contained in this book, we assume no responsibility for errors, inaccuracies, omissions, or any inconsistency herein. Any slighting of people, places, or organizations is unintentional.

First printing 2021
ISBN 978-1-7345880-8-8

How to Use this Book

This book is designed to help you learn your multiplication number facts. The goal is for you to have instant recall of each multiplication fact. Instant recall of number facts will make all math easier for you.

Here are some tips that will help you have the best possible learning experience.

- Be ready with at least 2 sharp pencils.
- Sit in a comfortable position so you can write quickly and neatly.
- It is good to have a timer so you can track your improvement in speed.
- For each speed circle, take the number in the center and multiply it with each number on the outside of the circle.
- At first, you might have to guess at some of the answers.
- **BIG TIP!** On page 139 is a multiplication table. Carefully remove it from the book and place it in a clear, protective, plastic sleeve. You will use this table to correct your work. Keep it in a safe place.
- When you correct your work be sure to make a check mark next to any mistake. Also, put the correct answer next to the check mark. This will help you to remember the correct answer.
- *Have fun!!* Track your accuracy as well as your speed.
- You will find that having instant recall of these facts will make all math easier for you.

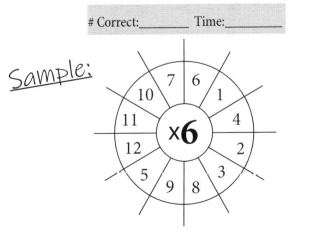

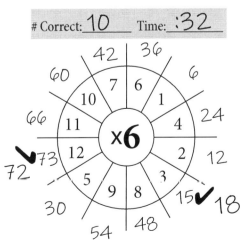

Don't miss the convenient Math Resource Center starting on page 127

21 Important "Math" Tips You Might Find Helpful

1. Learn your number facts. This makes all math so much easier.

2. Be neat. Use lined or graph paper to do your work. Sloppy work is often the cause of many mistakes. Always have at least 2 sharp pencils. Mechanical pencils are best.

3. Never do your math with a pen. This is living dangerously. We all make mistakes. Ink cannot be erased.

4. In a classroom or online, be sure to take notes and work problems right along with the instructor. That way you see it, you hear it, and you do it. You'll get the best learning experience if you do this.

5. Show steps in your work, especially when topics become more difficult.

6. If you fall behind, catch up as soon as possible. Waiting too long can make catching up a lot more difficult.

7. If you get a problem wrong, check to find out what caused the mistake. That way you won't make the same mistake again. Remember that everyone makes mistakes and we can all learn from them.

8. If there is something you don't understand, ask for help. Not later, but now.

9. Remember that everyone needs good math skills, and some of the most exciting careers require a strong background in math.

10. Success in algebra allows you to take more advanced math, science, and tech classes. Doing well in algebra should be an important goal for you. Work to become algebra-ready.

11. Find a "study buddy." Make sure it is someone you can work well with. You can help each other. Sometimes two heads are better than one.

12. Help someone else with their math. Teaching math to someone else is a good way to help you become a better math student.

13. Some schools have tutoring centers that can provide lots of help. Take advantage of them. They are a perfect place to get help and to do your math homework.

14. When you do your math homework at home, find a comfortable, quiet place to work. This is the best way to avoid distractions.

15. Sometimes it is helpful to use a diagram, a chart, a graph, or a sketch to solve math problems.

16. When studying example problems that show steps, don't just read them. Copy the example problem neatly and write out all the steps. There is something about writing them down that will help you to understand them better, as well as remember them.

17. Be sure to review topics that you have learned. This will help you retain what you have learned about past topics. You won't forget them

18. Remember that we all make mistakes. We all struggle at times. Don't give up! Never give up! That's all part of life.

19. Connect the learning of math with a long-term goal. For example, if you want to become a doctor, you will need to take lots of math and science classes to achieve this goal.

20. No excuses! Take advantage of the many sources for help. These include teachers, tutors, parents, siblings, friends, and online courses. There is lots of help available. Go for it!

21. If you have a career goal, make a connection with someone involved in that career and set up a time to interview them. Make a list of questions and get as much advice as possible about achieving that career goal. Be sure to ask about the importance of math!

Speed Wheel Drills for Multiplication

7

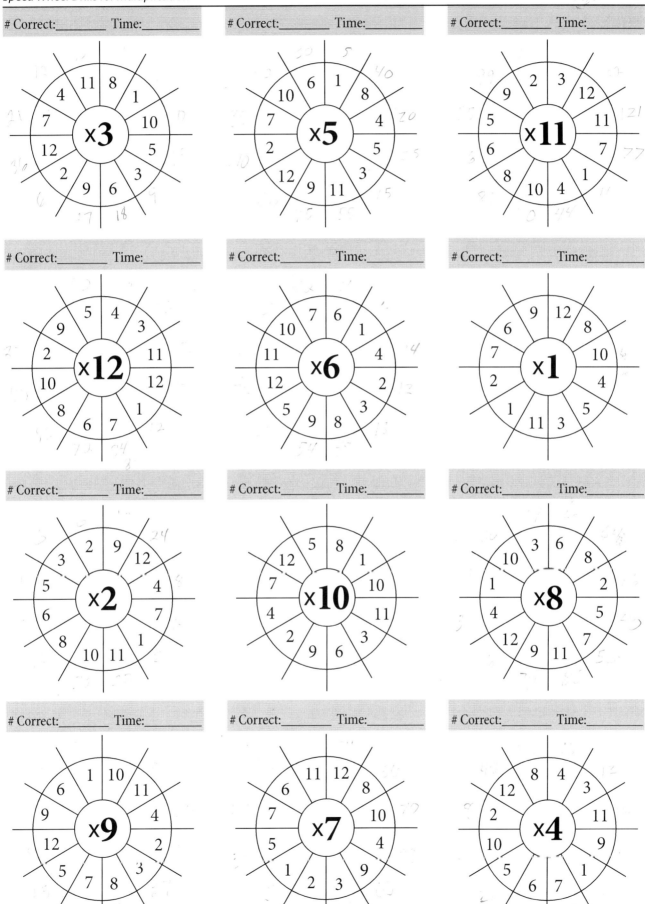

It is illegal to photocopy this page. Copyright © 2021, Richard W. Fisher

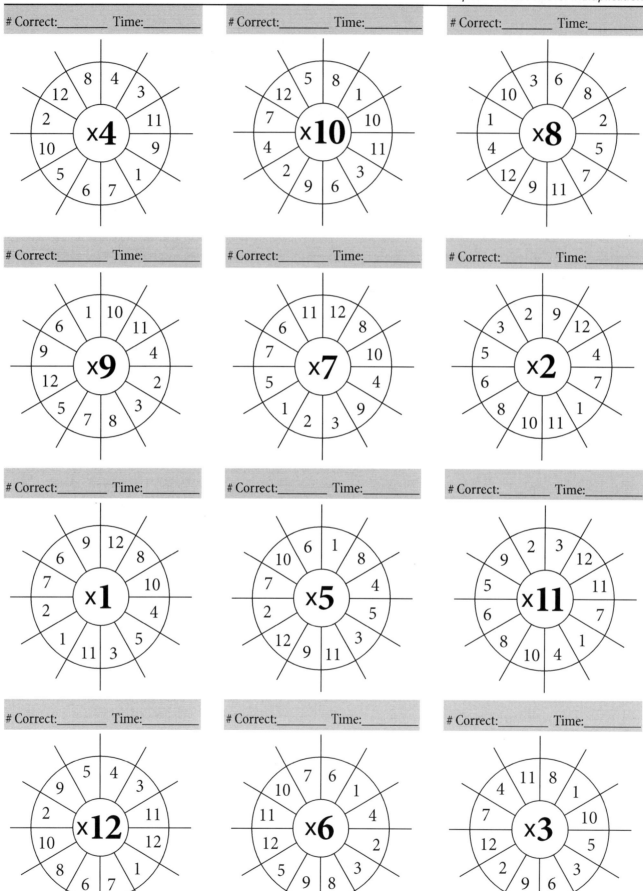

Speed Wheel Drills for Multiplication 9

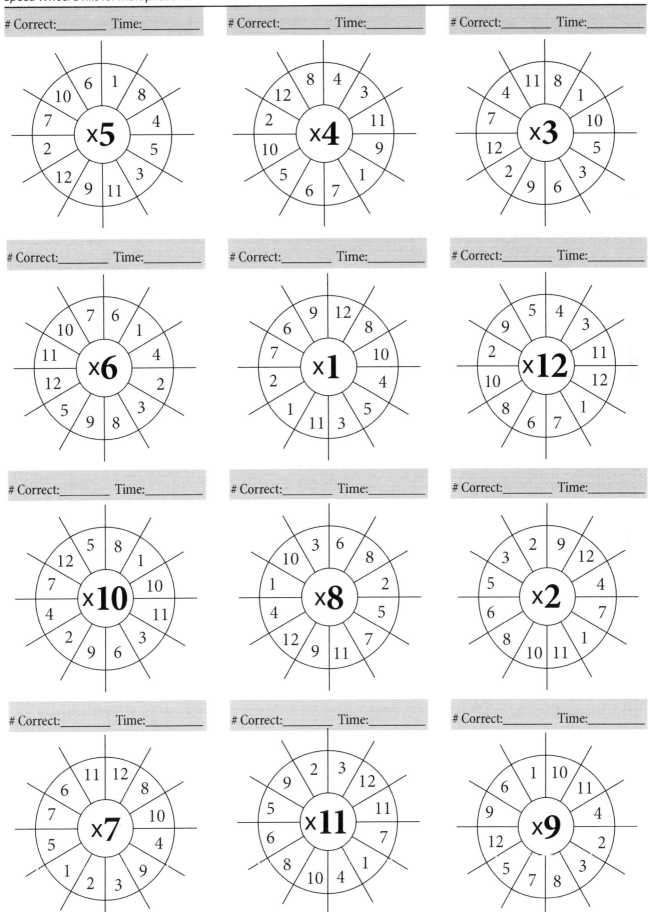

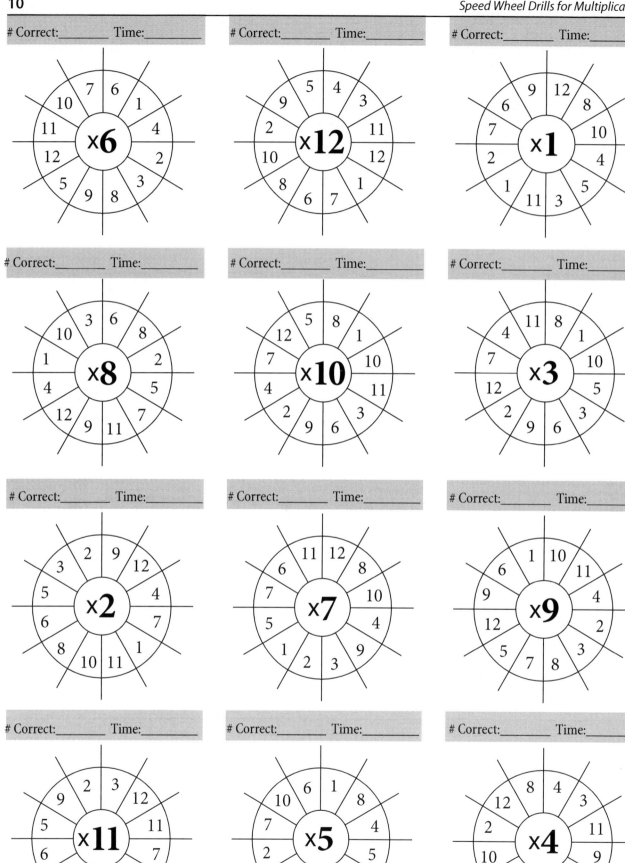

Speed Wheel Drills for Multiplication 11

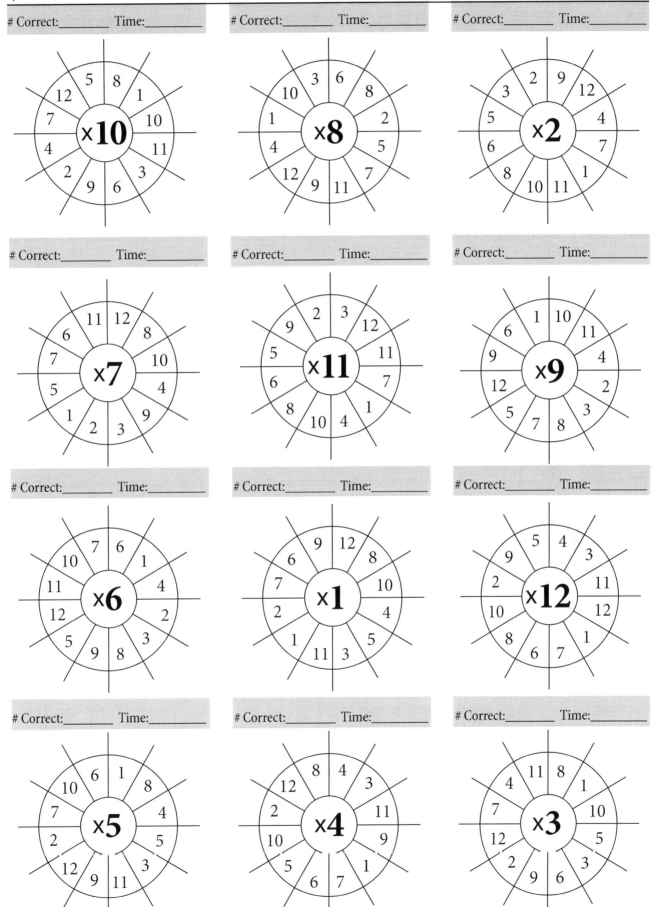

12 Speed Wheel Drills for Multiplication

Correct:_____ Time:_____

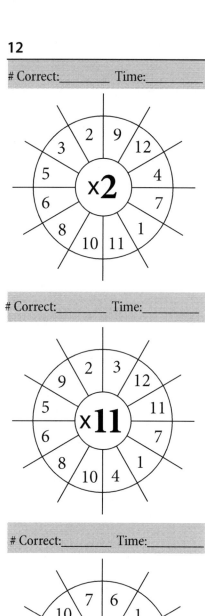

Correct:_____ Time:_____

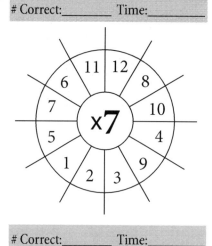

Correct:_____ Time:_____

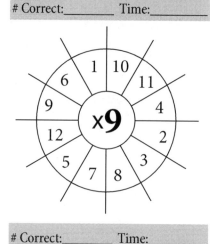

Correct:_____ Time:_____

Correct:_____ Time:_____

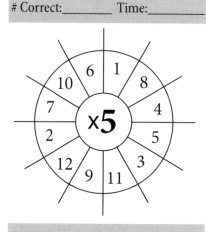

Correct:_____ Time:_____

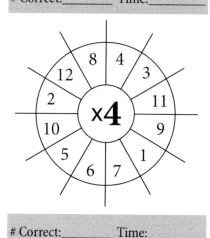

Correct:_____ Time:_____

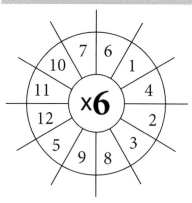

Correct:_____ Time:_____

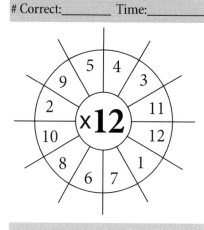

Correct:_____ Time:_____

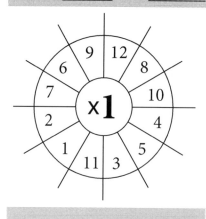

Correct:_____ Time:_____

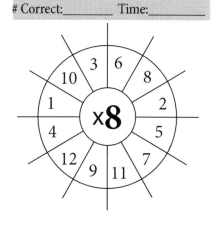

Correct:_____ Time:_____

Correct:_____ Time:_____

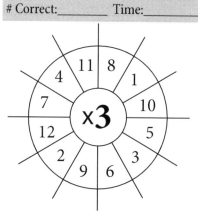

It is illegal to photocopy this page. Copyright © 2021, Richard W. Fisher

Speed Wheel Drills for Multiplication 13

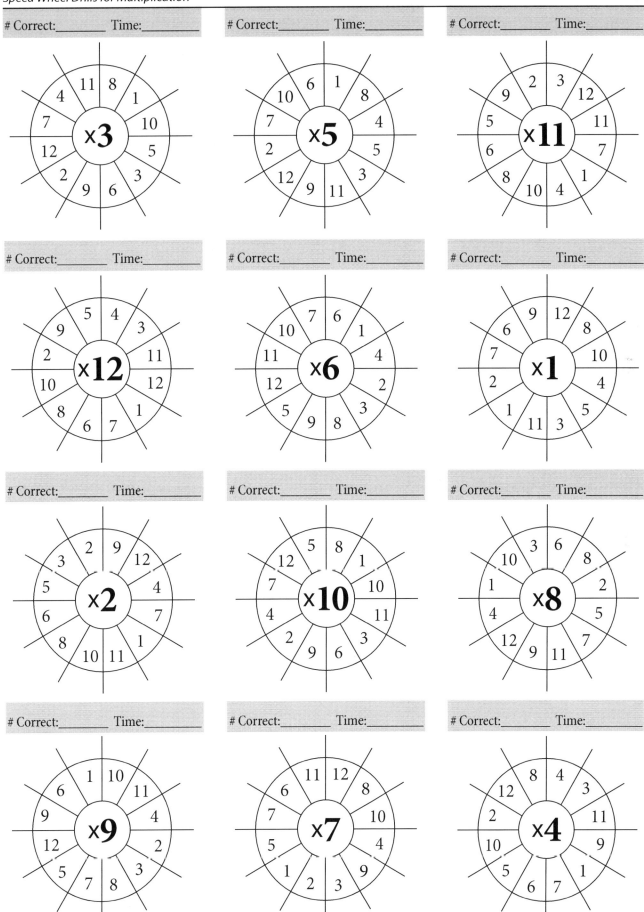

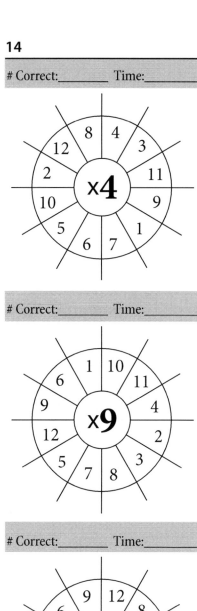
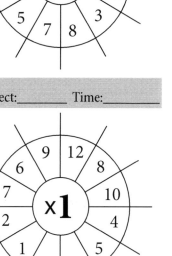
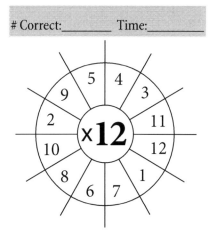
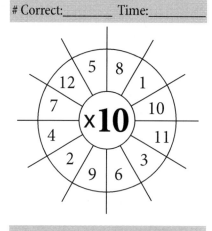
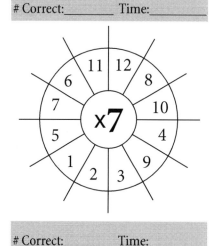
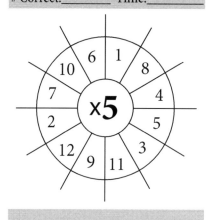
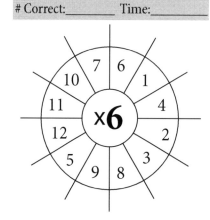
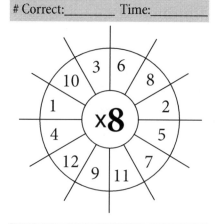
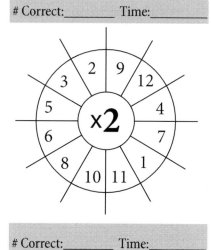

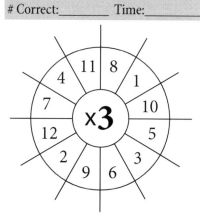

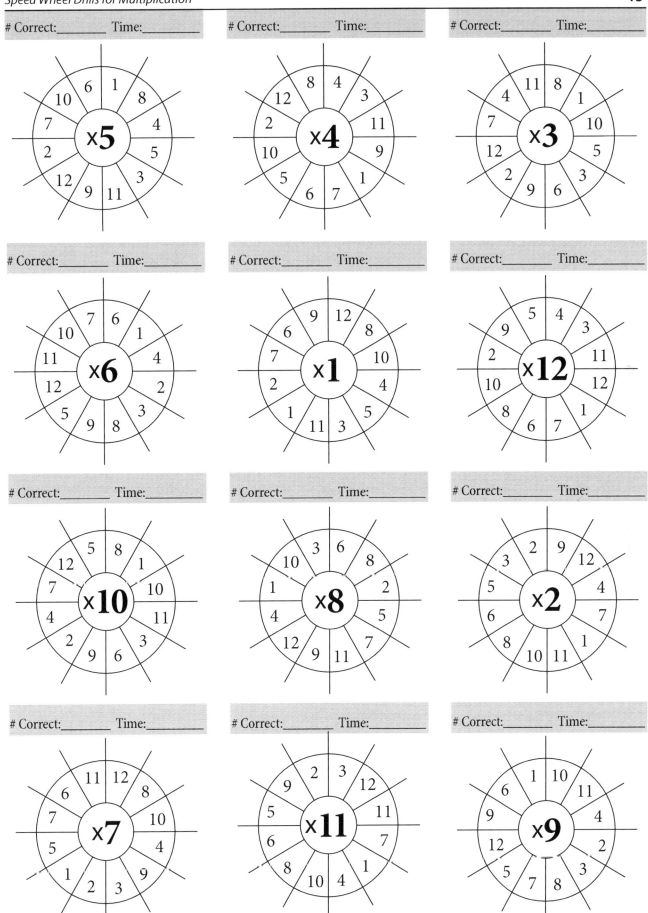

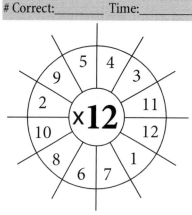
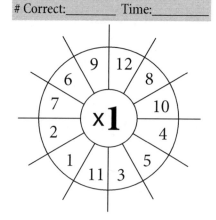
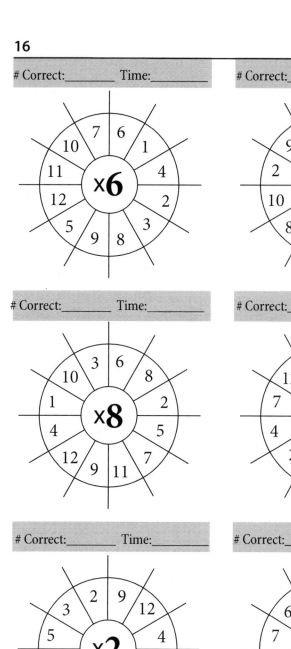

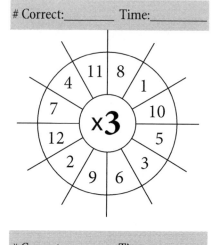
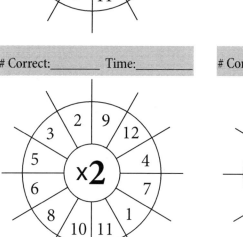
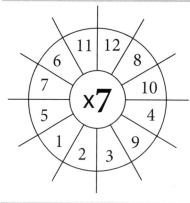

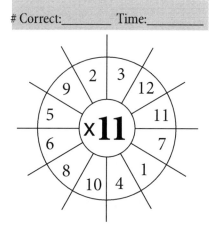
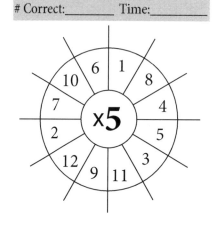
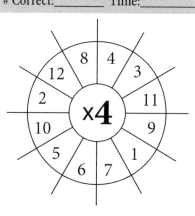

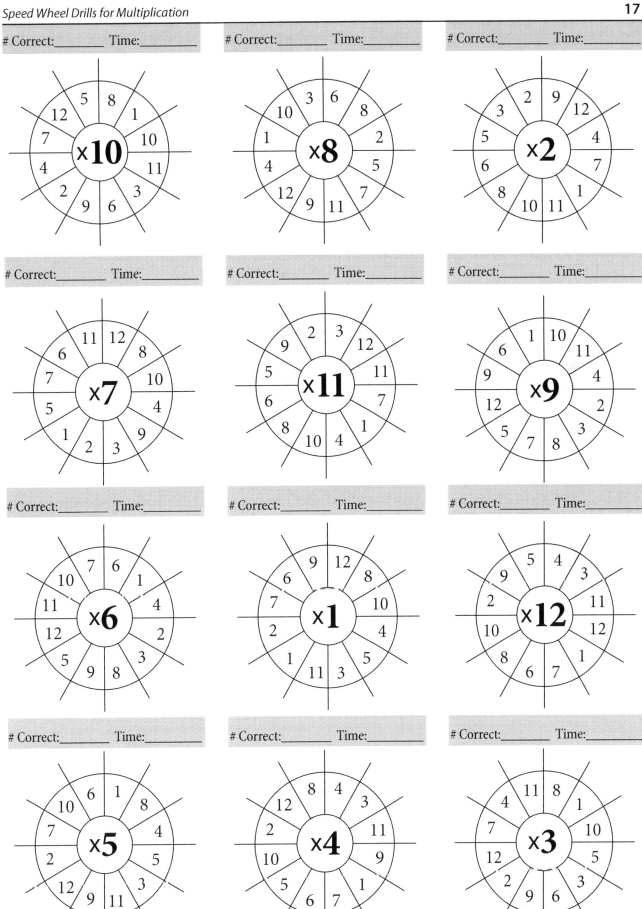

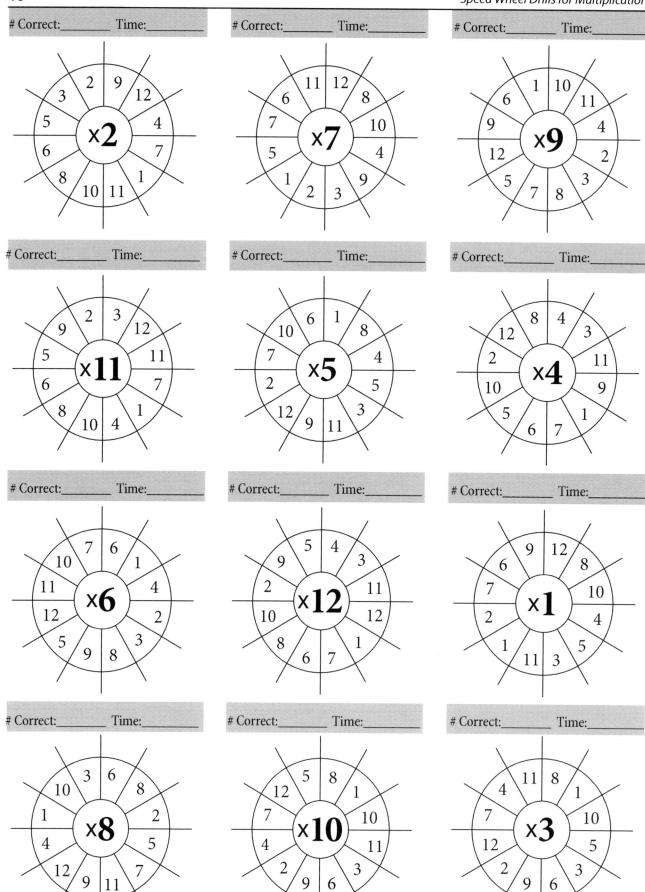

Speed Wheel Drills for Multiplication 19

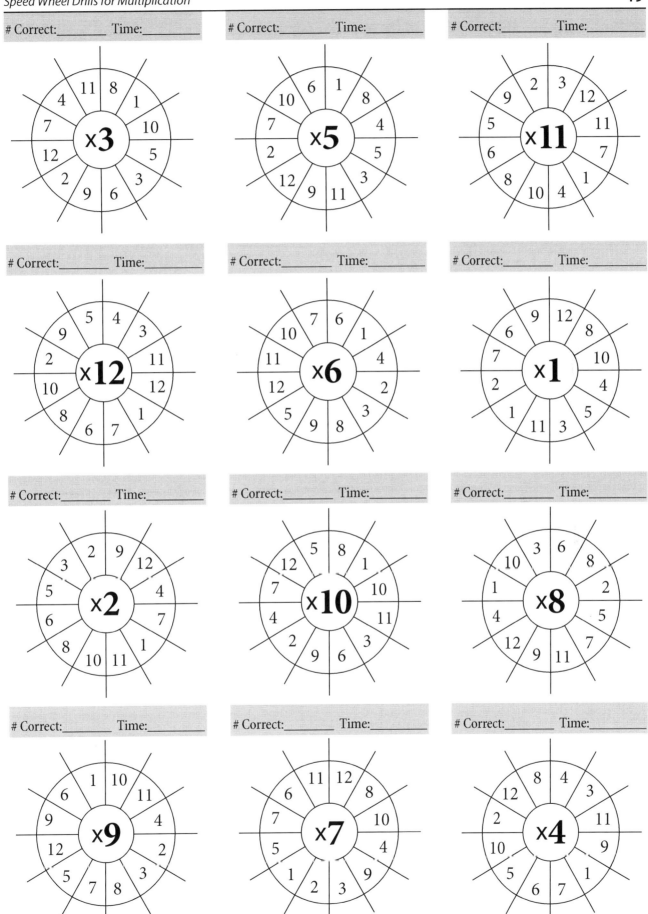

Correct:_____ Time:_____

Correct:_____ Time:_____

Correct:_____ Time:_____

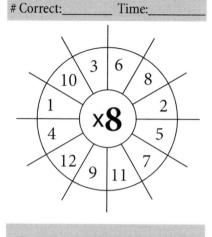

Correct:_____ Time:_____

Correct:_____ Time:_____

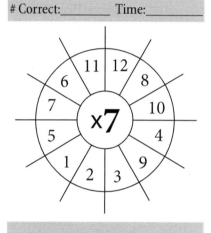

Correct:_____ Time:_____

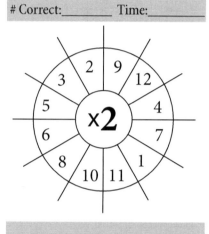

Correct:_____ Time:_____

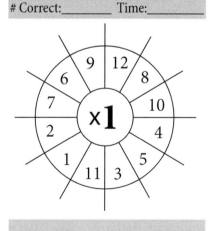

Correct:_____ Time:_____

Correct:_____ Time:_____

Correct:_____ Time:_____

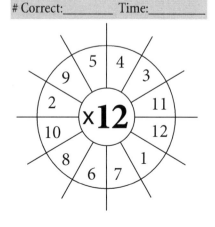

Correct:_____ Time:_____

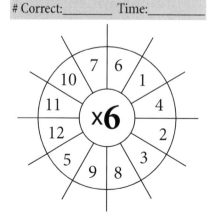

Correct:_____ Time:_____

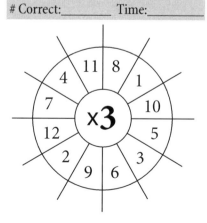

Speed Wheel Drills for Multiplication 21

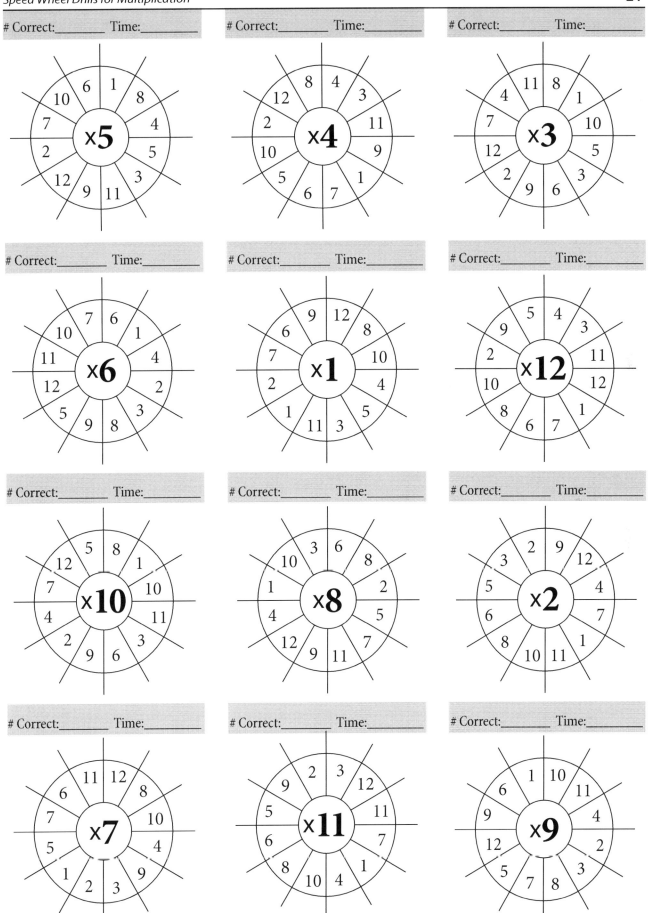

Correct:_____ Time:_____
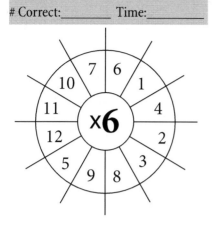

Correct:_____ Time:_____
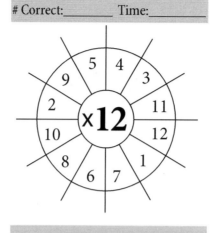

Correct:_____ Time:_____
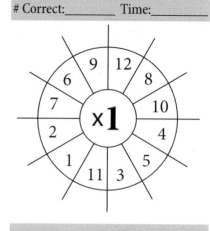

Correct:_____ Time:_____
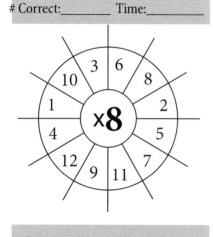

Correct:_____ Time:_____
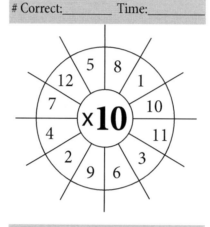

Correct:_____ Time:_____
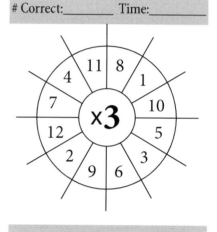

Correct:_____ Time:_____
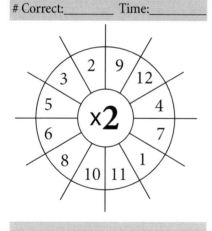

Correct:_____ Time:_____
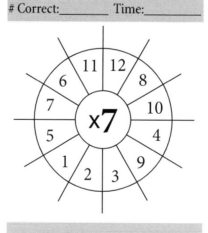

Correct:_____ Time:_____

Correct:_____ Time:_____
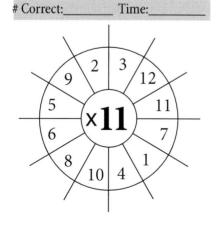

Correct:_____ Time:_____
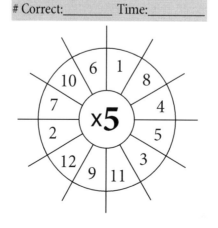

Correct:_____ Time:_____
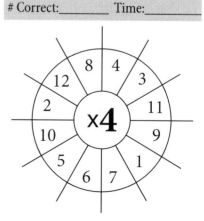

Speed Wheel Drills for Multiplication 23

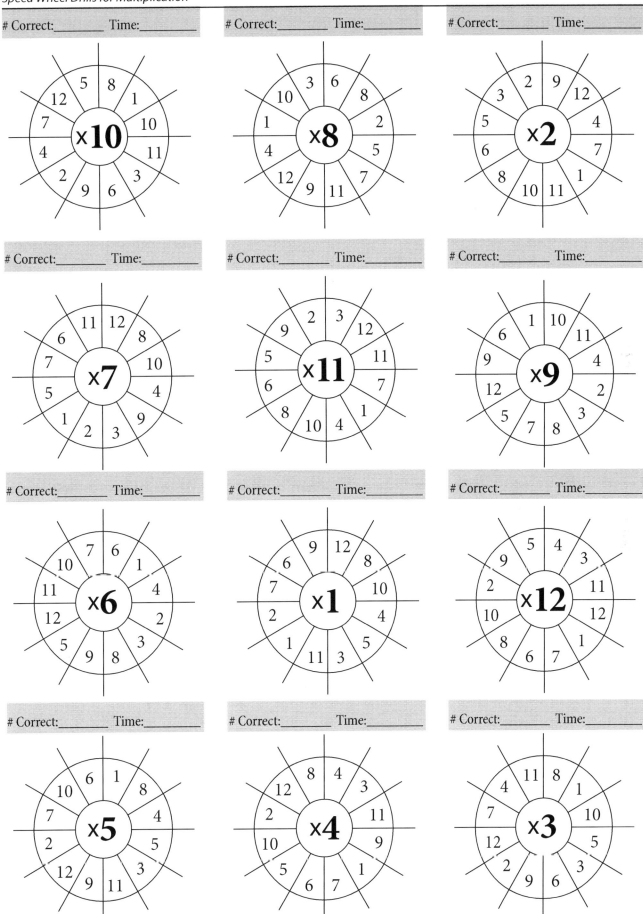

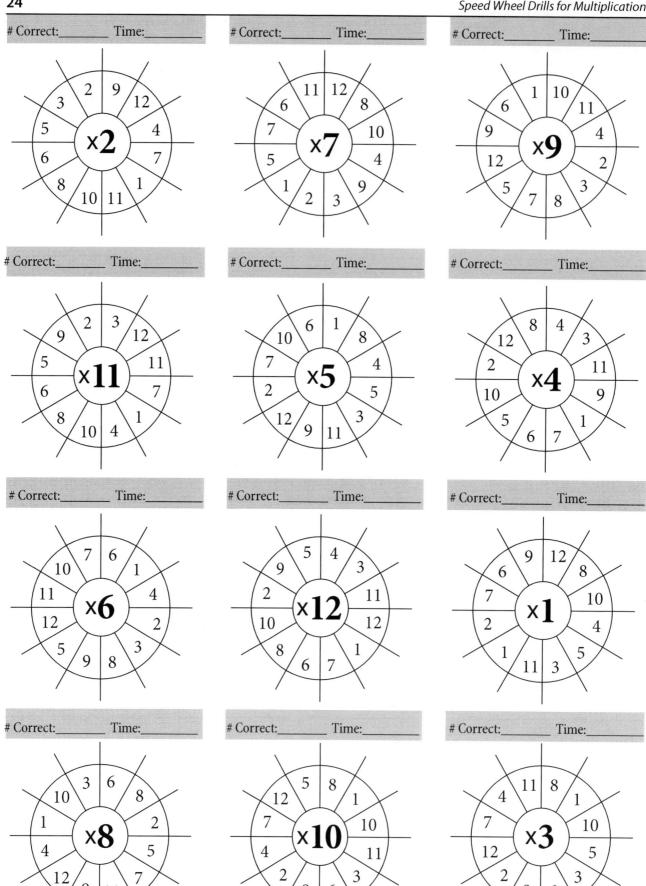

Speed Wheel Drills for Multiplication

25

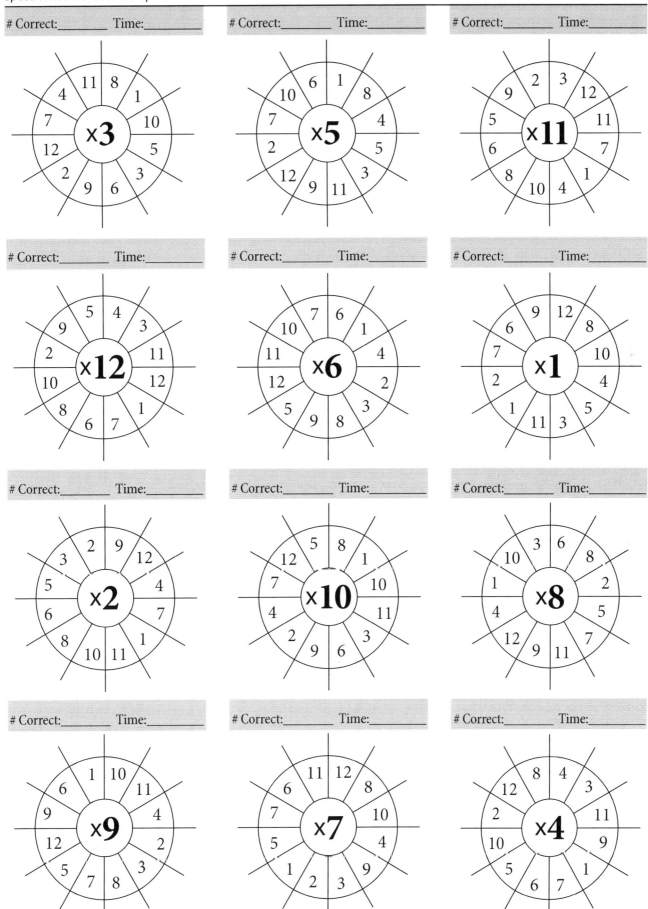

It is illegal to photocopy this page. Copyright © 2021, Richard W. Fisher

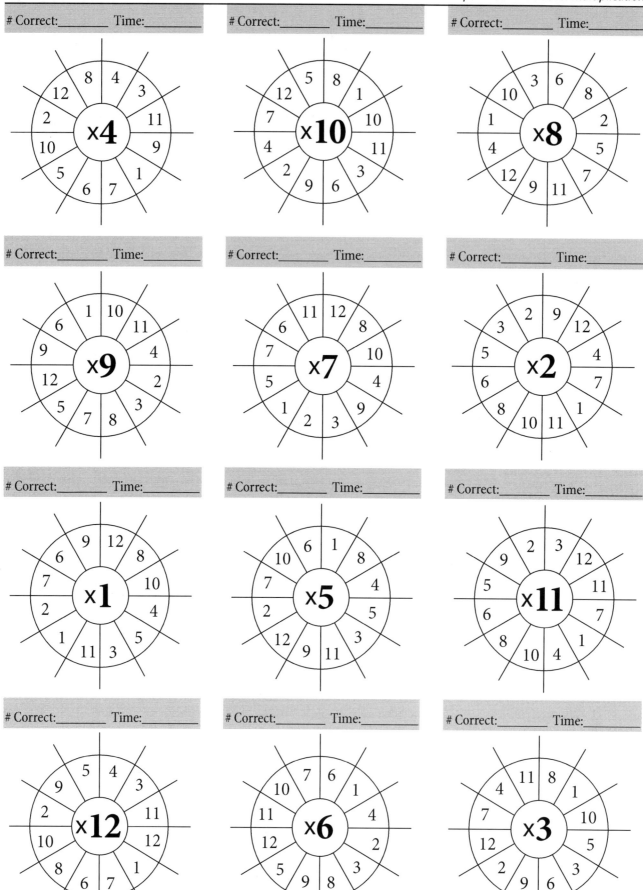

Speed Wheel Drills for Multiplication 27

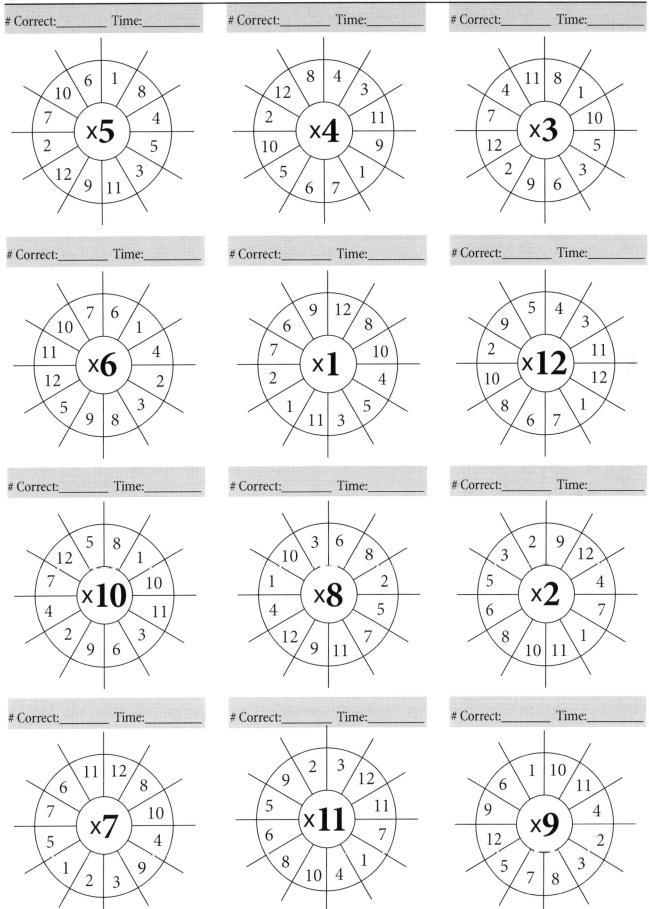

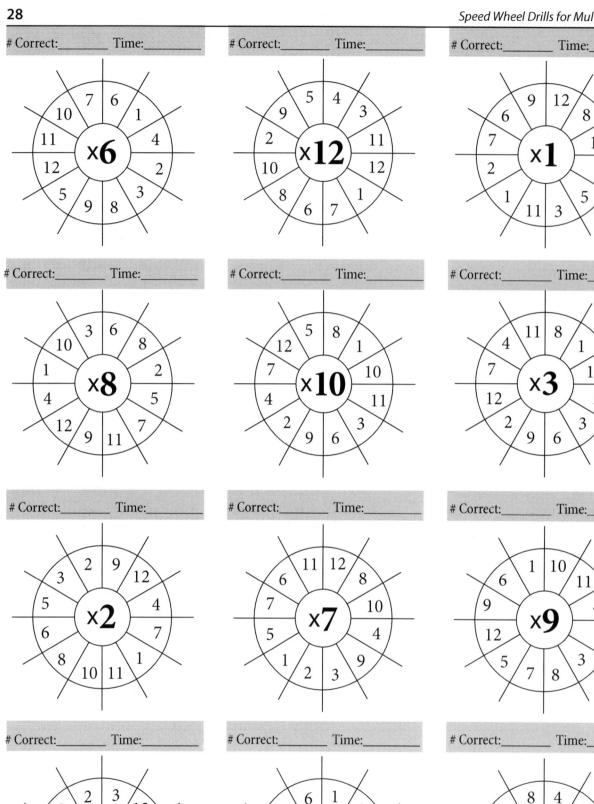

Speed Wheel Drills for Multiplication

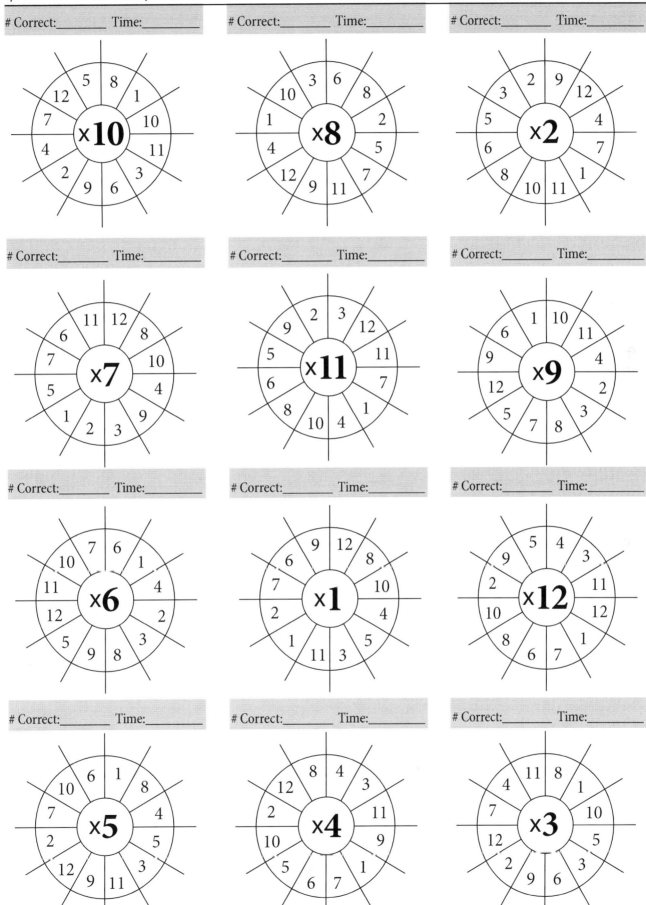

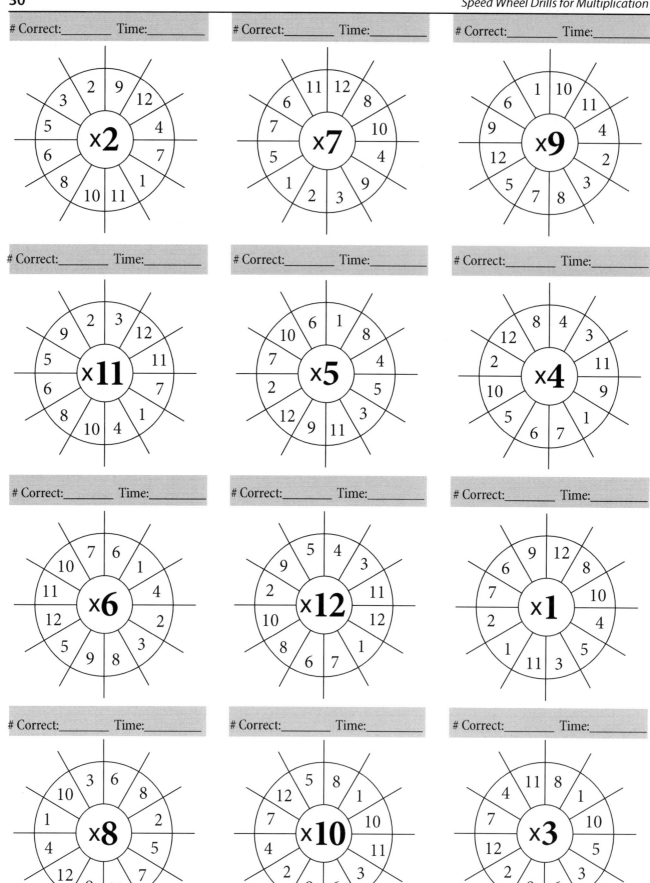

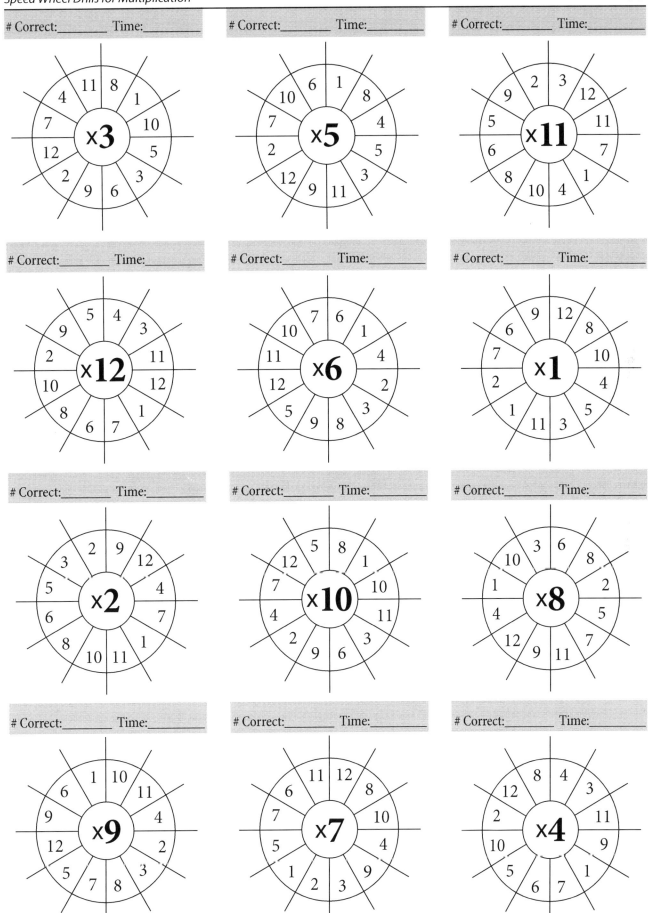

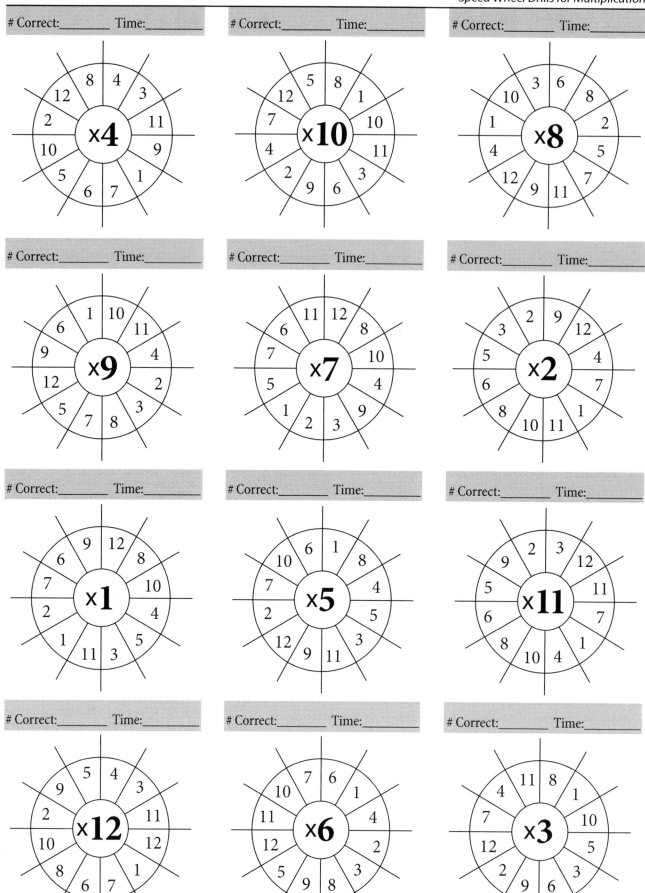

Speed Wheel Drills for Multiplication 33

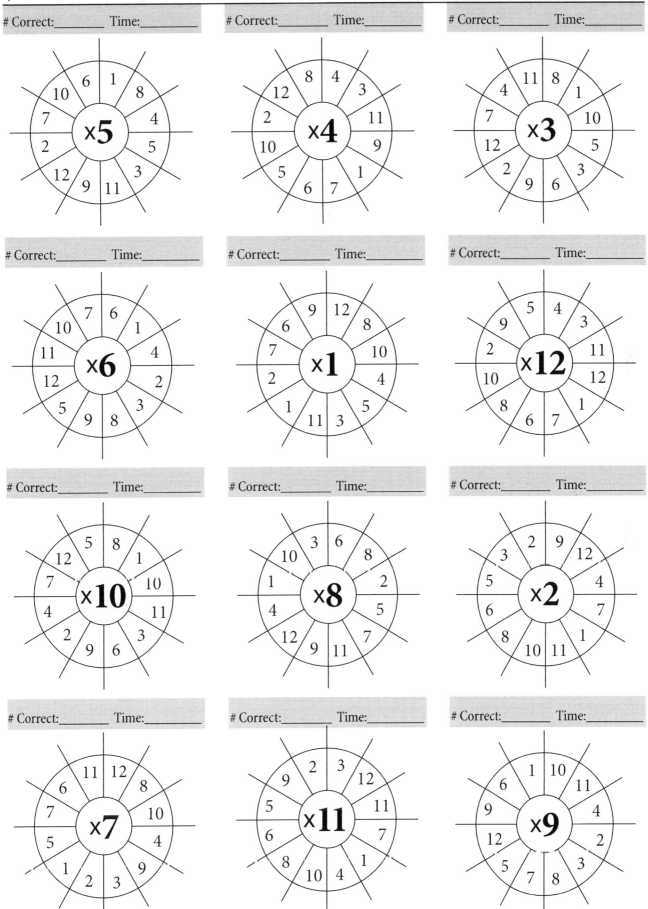

Correct:_____ Time:_____   # Correct:_____ Time:_____   # Correct:_____ Time:_____

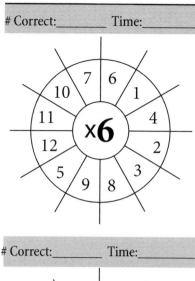  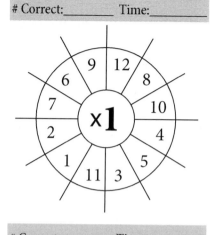

Correct:_____ Time:_____   # Correct:_____ Time:_____   # Correct:_____ Time:_____

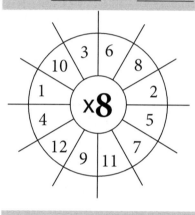 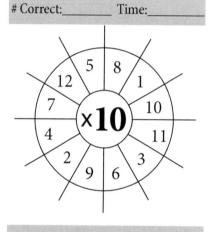 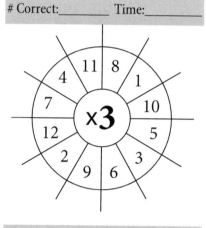

Correct:_____ Time:_____   # Correct:_____ Time:_____   # Correct:_____ Time:_____

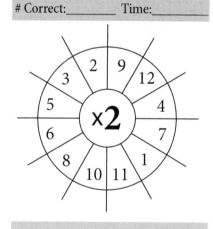 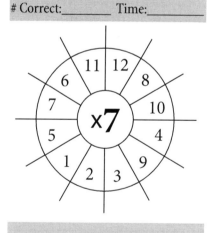 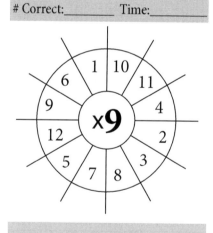

Correct:_____ Time:_____   # Correct:_____ Time:_____   # Correct:_____ Time:_____

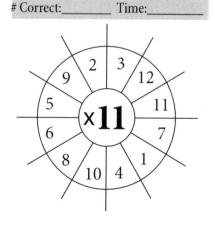 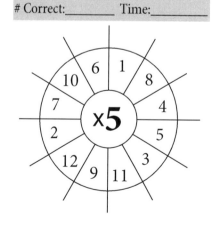 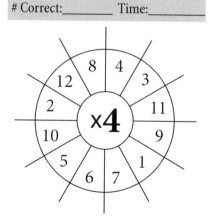

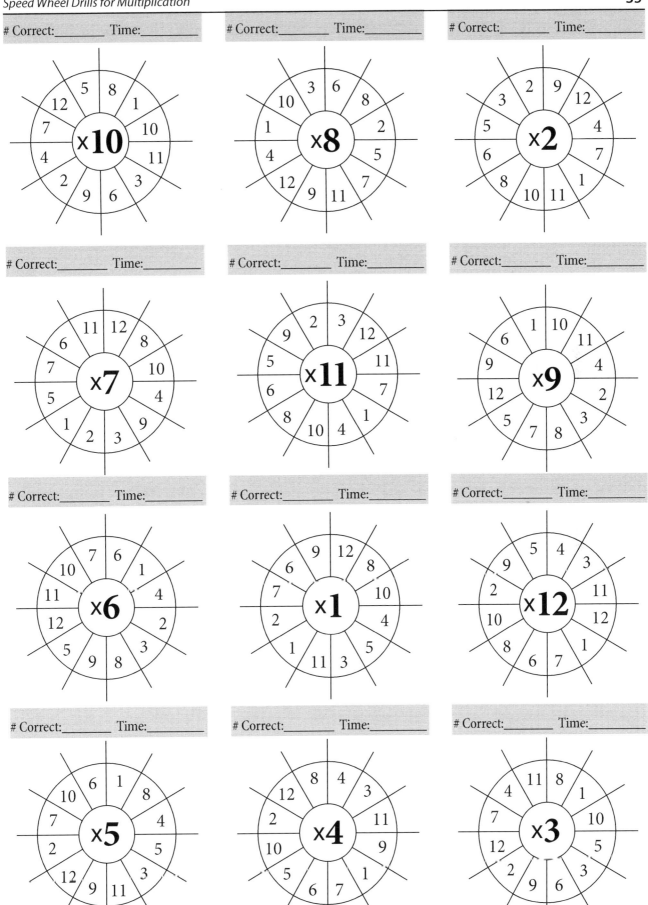

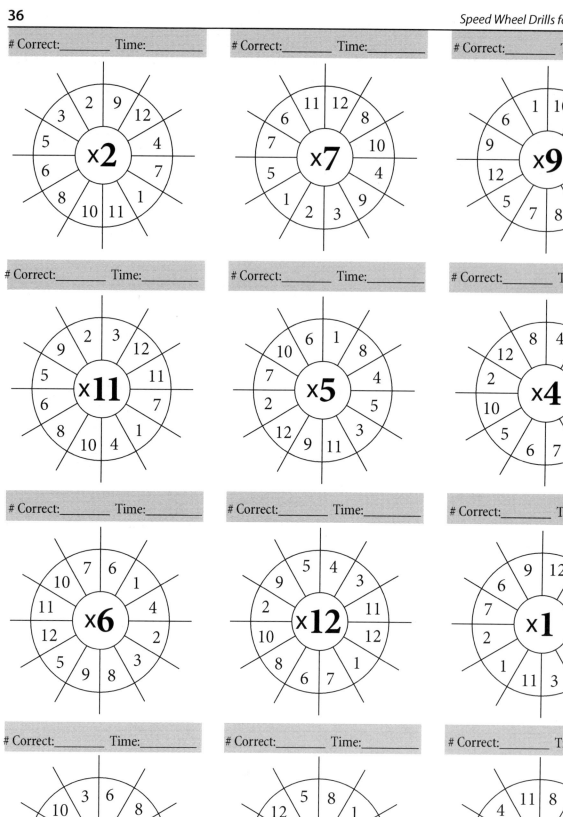

Speed Wheel Drills for Multiplication 37

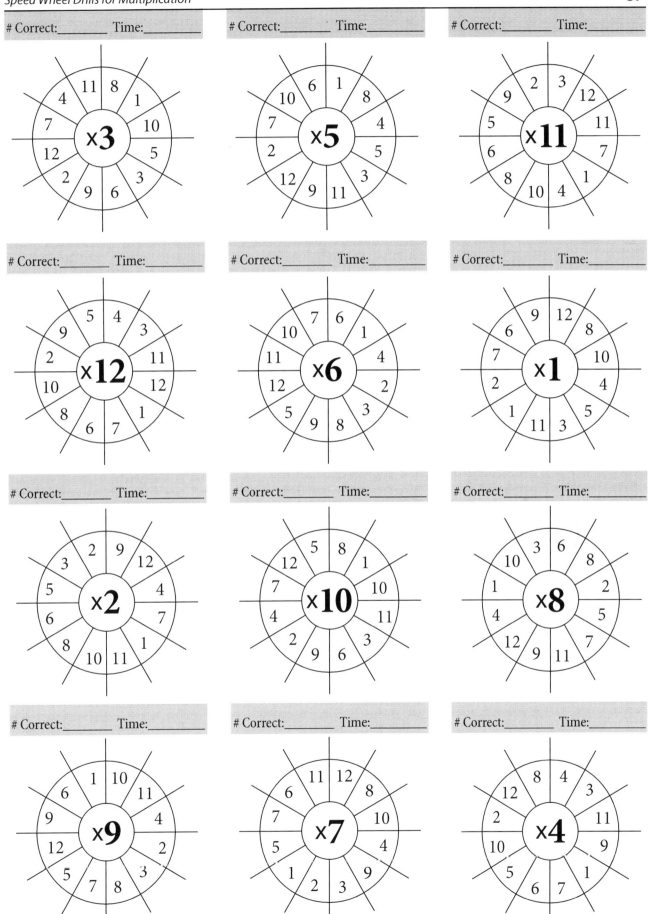

It is illegal to photocopy this page. Copyright © 2021, Richard W. Fisher

Correct:_____ Time:_____   # Correct:_____ Time:_____   # Correct:_____ Time:_____

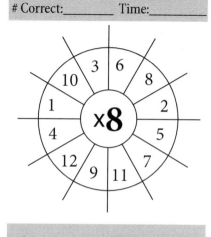

Correct:_____ Time:_____   # Correct:_____ Time:_____   # Correct:_____ Time:_____

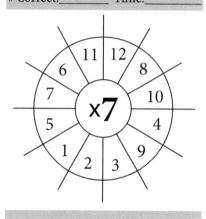
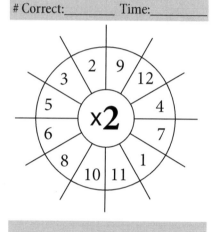

Correct:_____ Time:_____   # Correct:_____ Time:_____   # Correct:_____ Time:_____

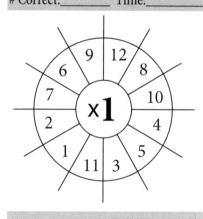
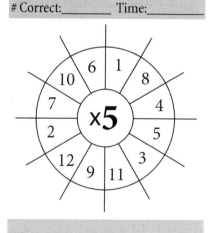

Correct:_____ Time:_____   # Correct:_____ Time:_____   # Correct:_____ Time:_____

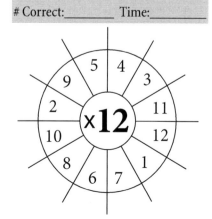
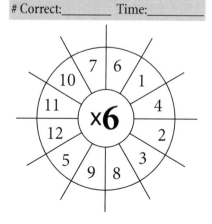
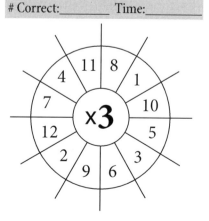

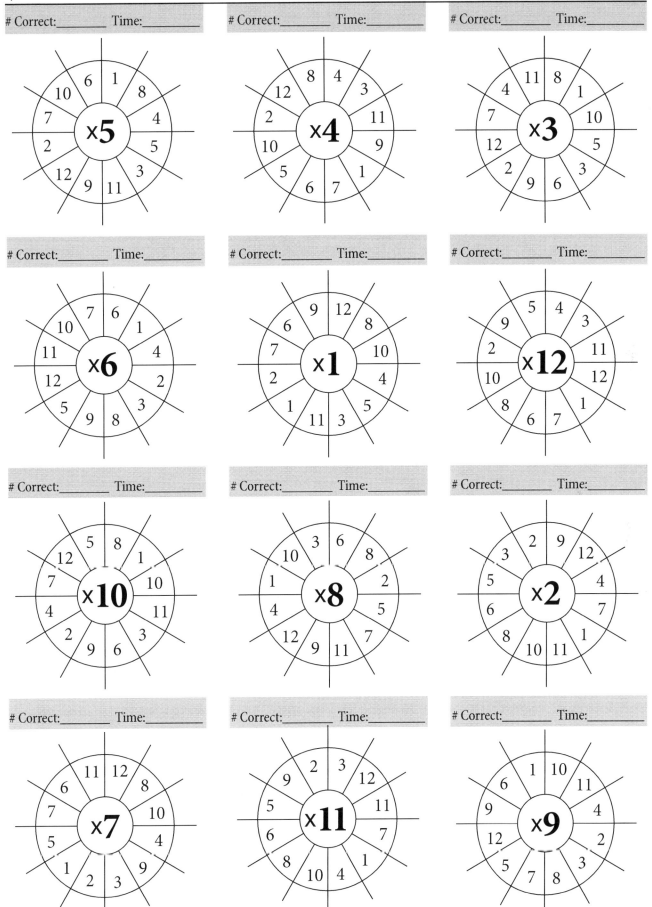

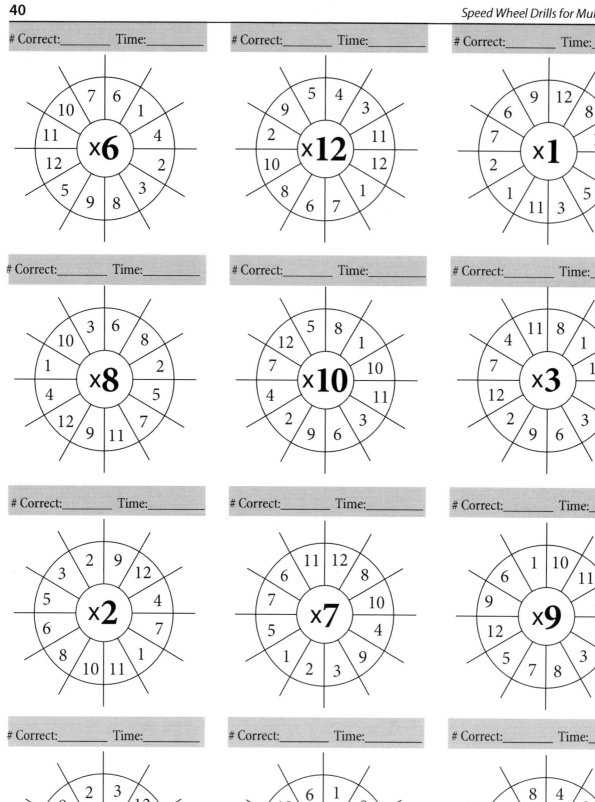

Speed Wheel Drills for Multiplication 41

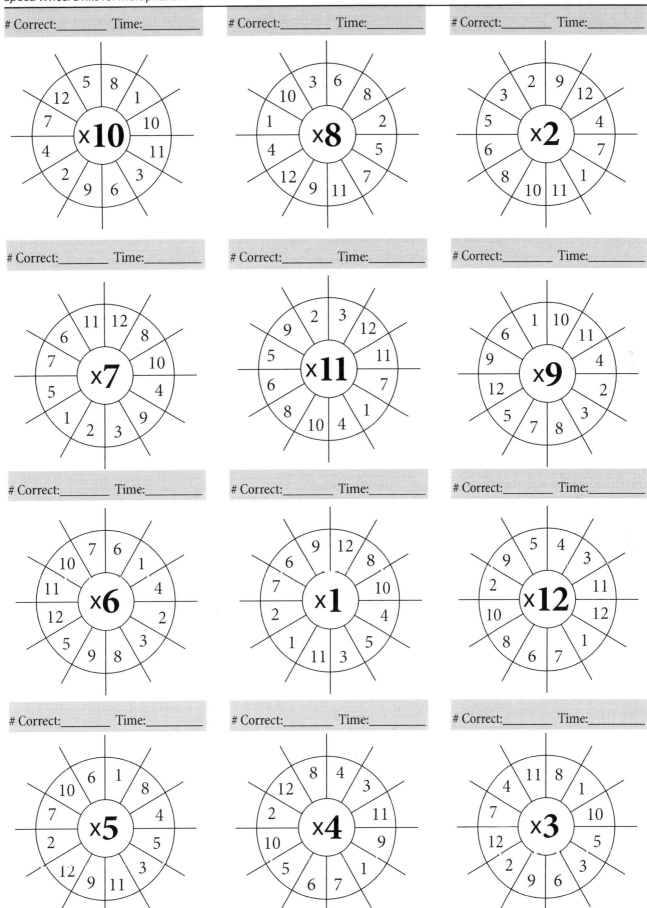

Speed Wheel Drills for Multiplication

Correct:_____ Time:_____
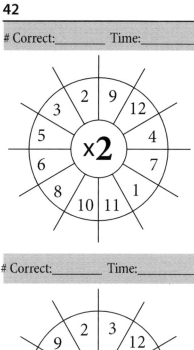

Correct:_____ Time:_____
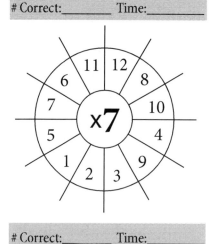

Correct:_____ Time:_____
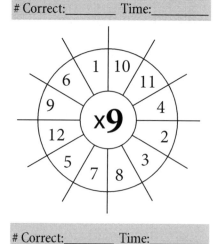

Correct:_____ Time:_____
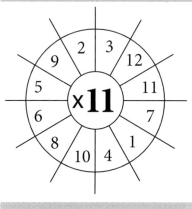

Correct:_____ Time:_____
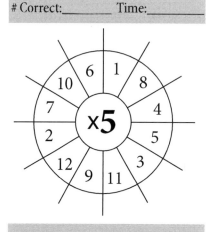

Correct:_____ Time:_____

Correct:_____ Time:_____
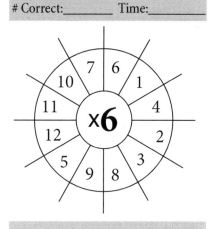

Correct:_____ Time:_____

Correct:_____ Time:_____
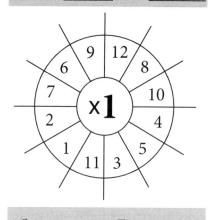

Correct:_____ Time:_____
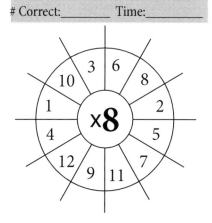

Correct:_____ Time:_____
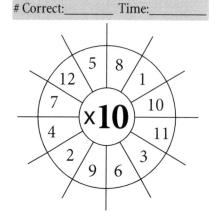

Correct:_____ Time:_____
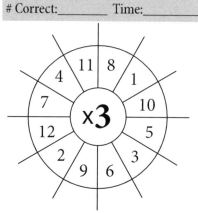

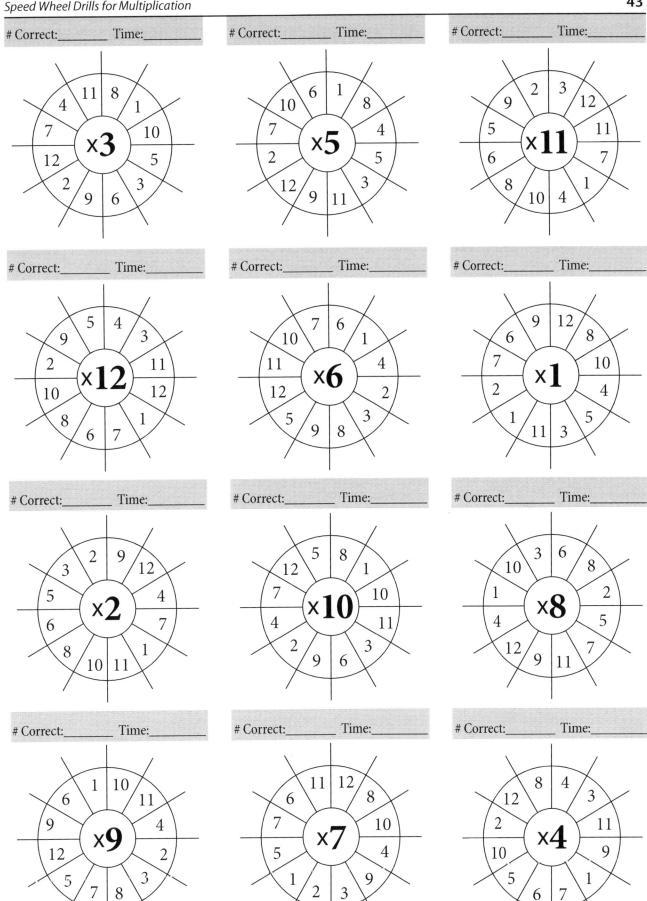

44 — Speed Wheel Drills for Multiplication

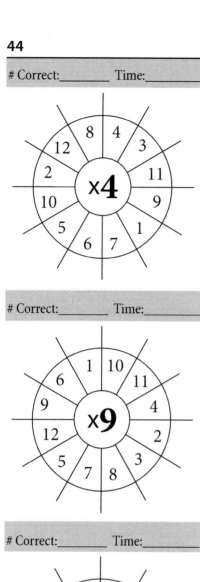
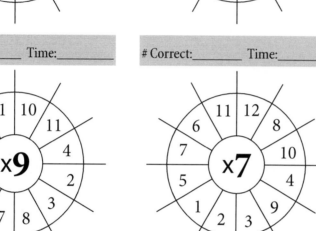
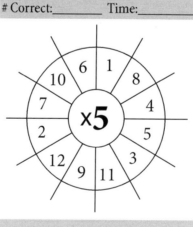
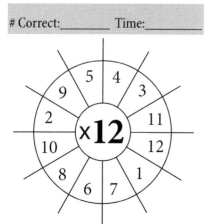
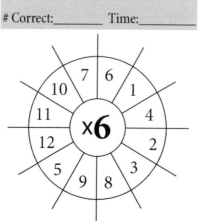
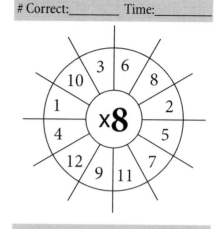
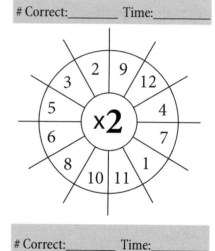

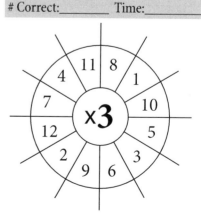

Speed Wheel Drills for Multiplication 45

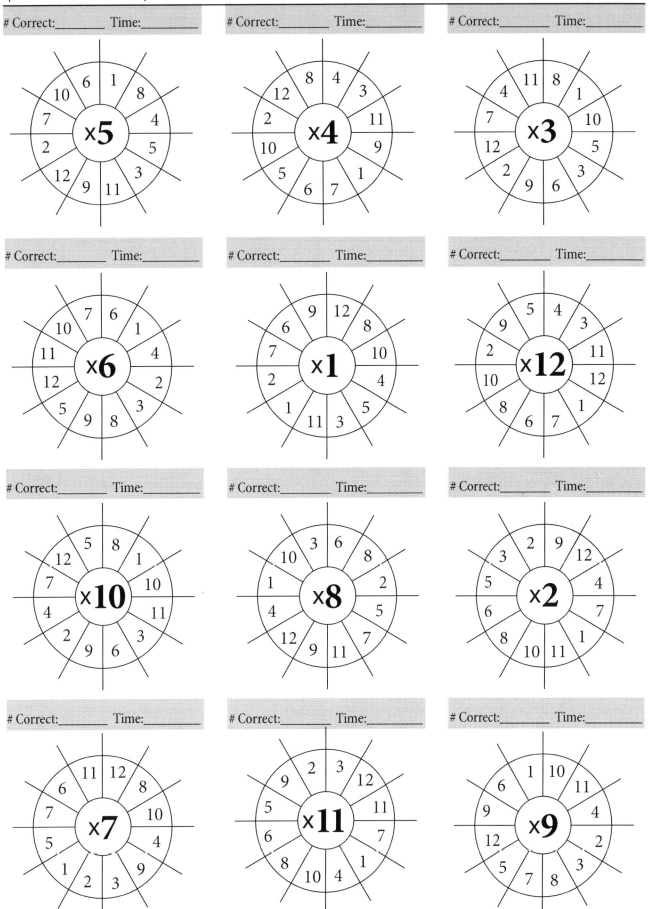

Correct:_____ Time:_____   # Correct:_____ Time:_____   # Correct:_____ Time:_____

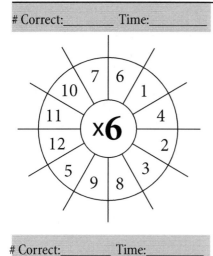 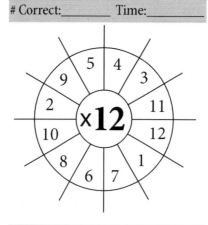 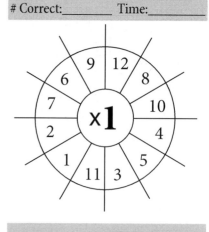

Correct:_____ Time:_____   # Correct:_____ Time:_____   # Correct:_____ Time:_____

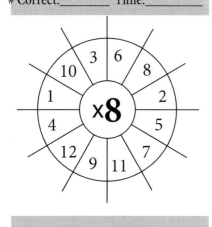 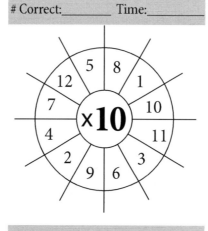 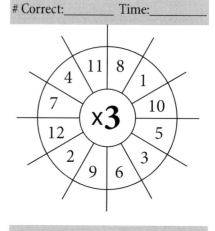

Correct:_____ Time:_____   # Correct:_____ Time:_____   # Correct:_____ Time:_____

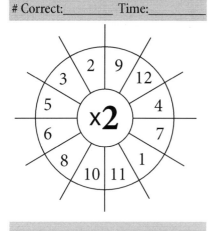 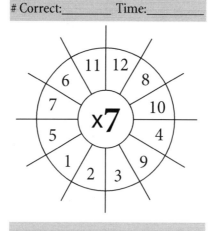 

Correct:_____ Time:_____   # Correct:_____ Time:_____   # Correct:_____ Time:_____

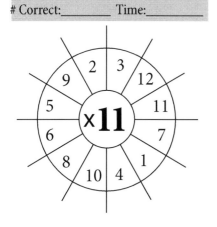 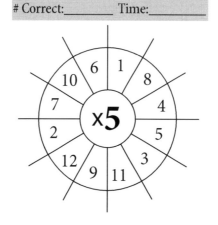 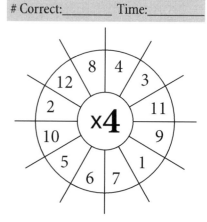

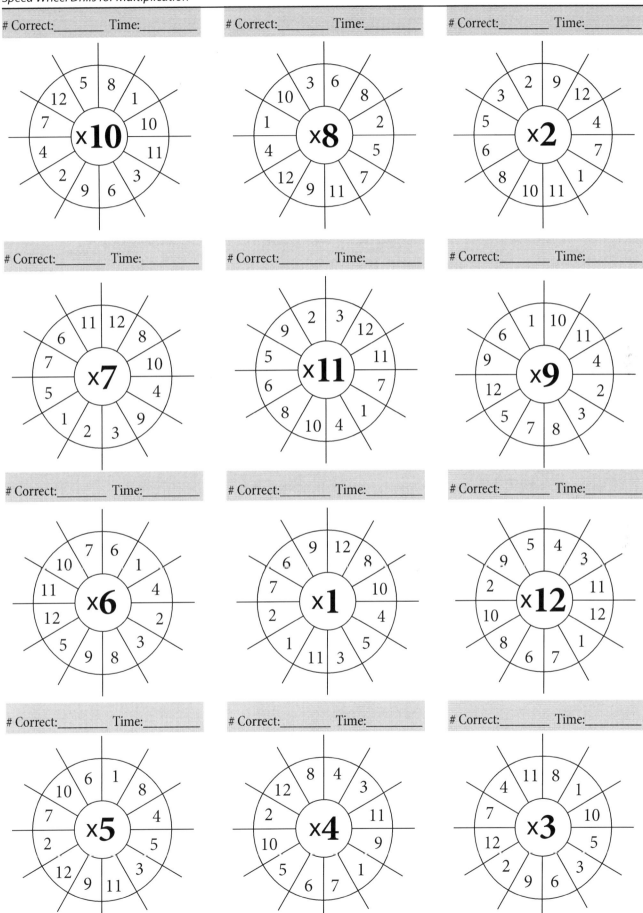

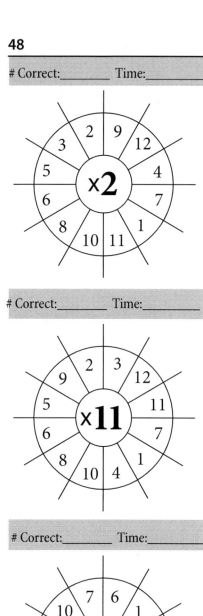
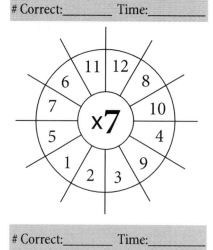
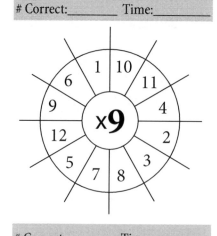
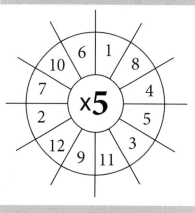
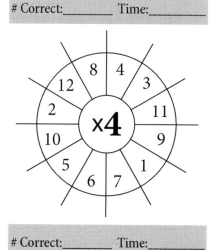
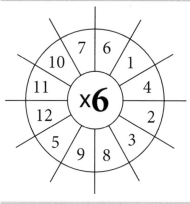

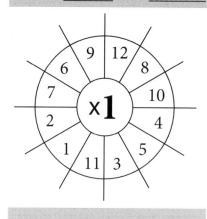
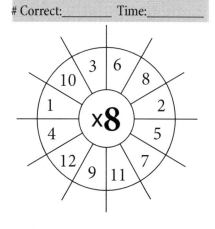
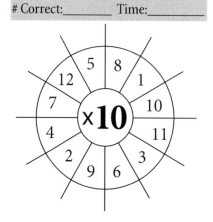
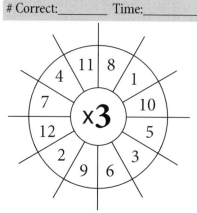

Speed Wheel Drills for Multiplication 49

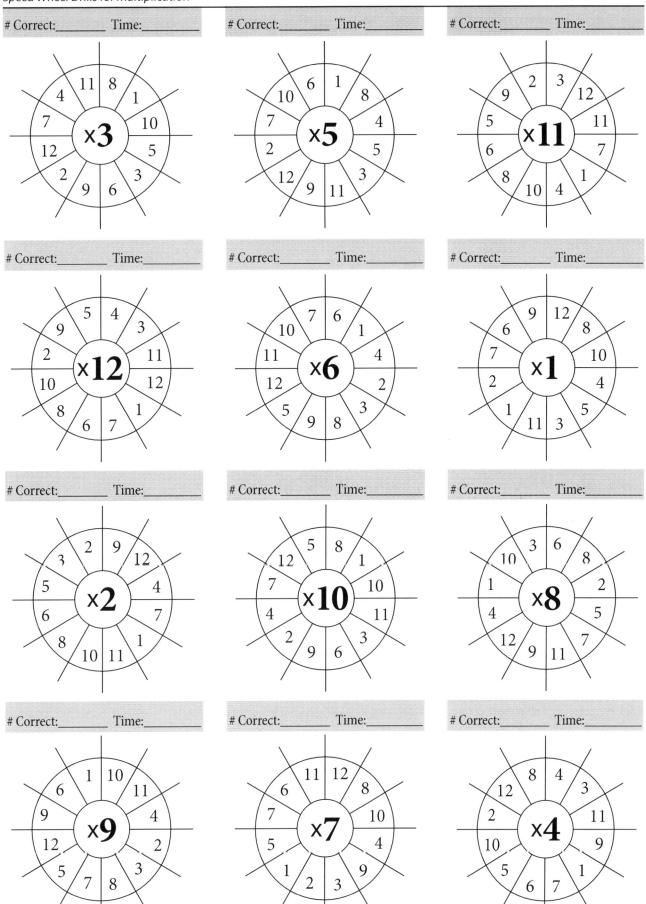

50

Speed Wheel Drills for Multiplication

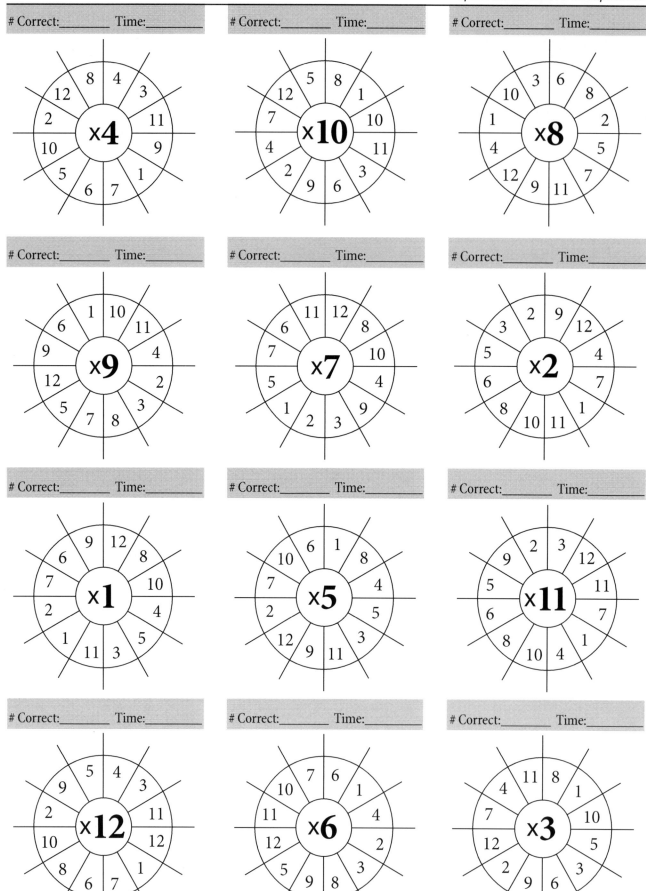

It is illegal to photocopy this page. Copyright © 2021, Richard W. Fisher

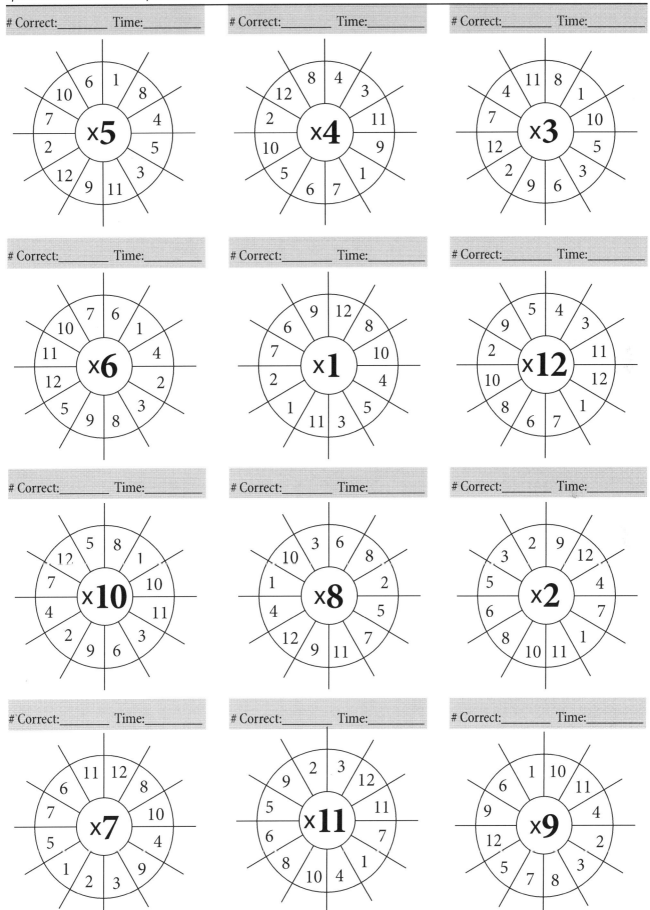

Speed Wheel Drills for Multiplication

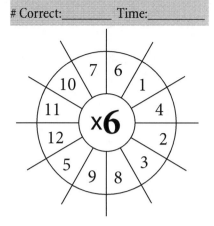
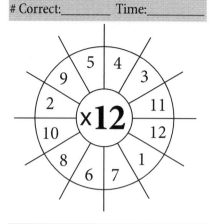
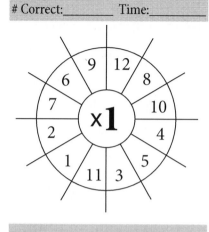
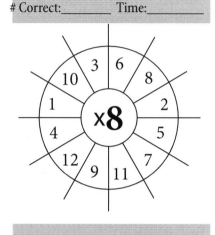
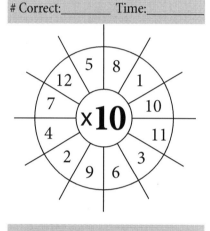
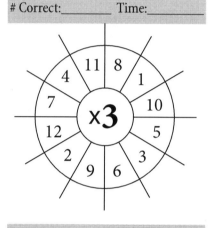
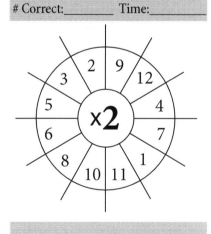
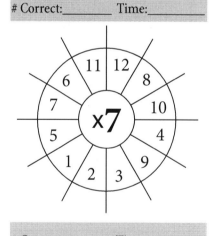
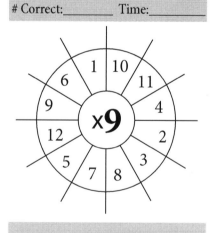
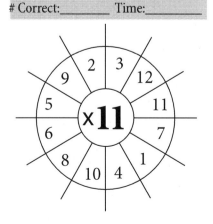
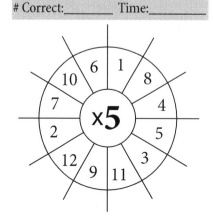
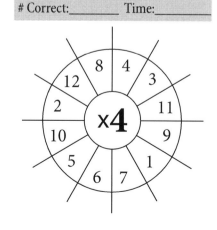

It is illegal to photocopy this page. Copyright © 2021, Richard W. Fisher

Speed Wheel Drills for Multiplication

53

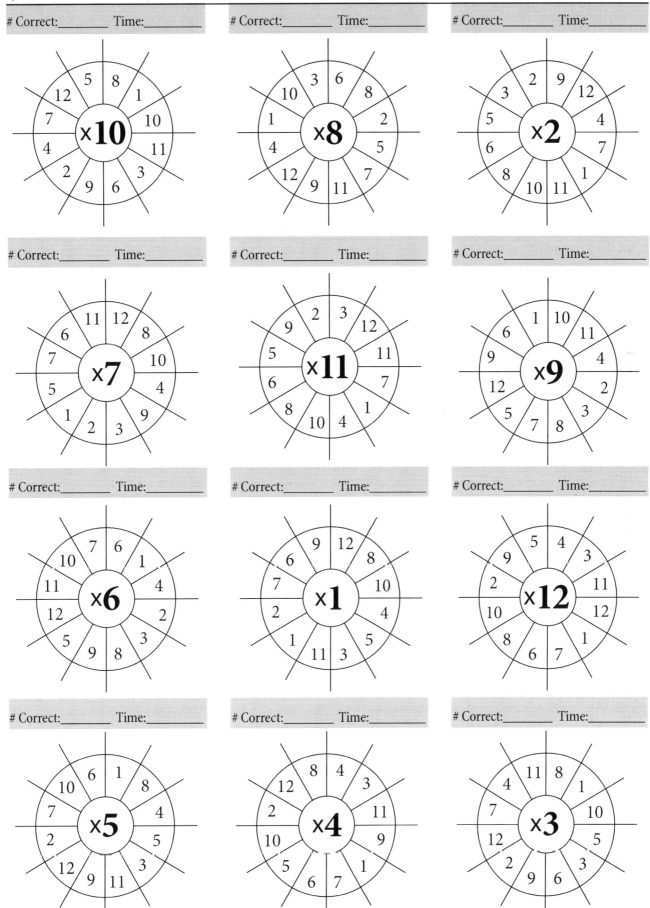

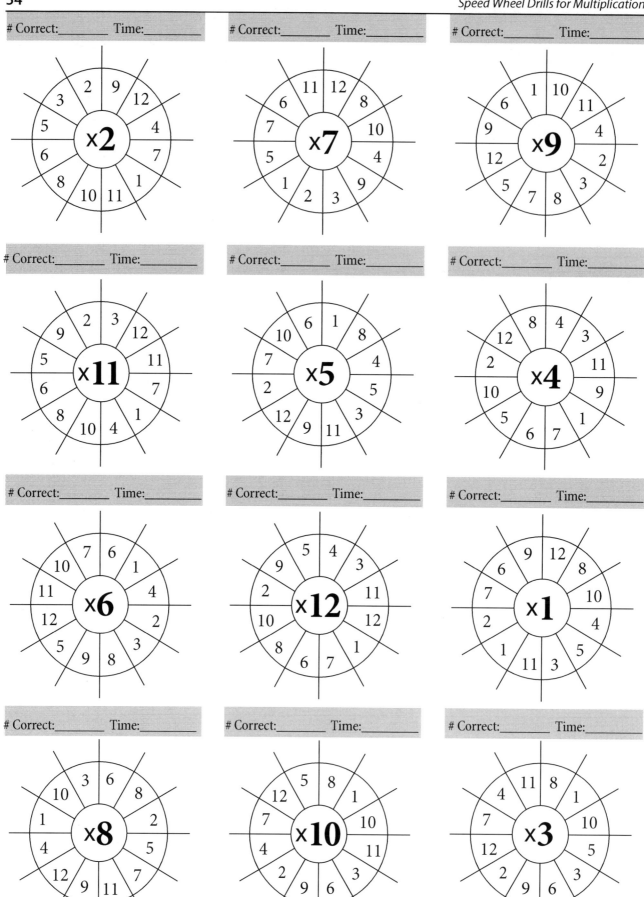

Speed Wheel Drills for Multiplication 55

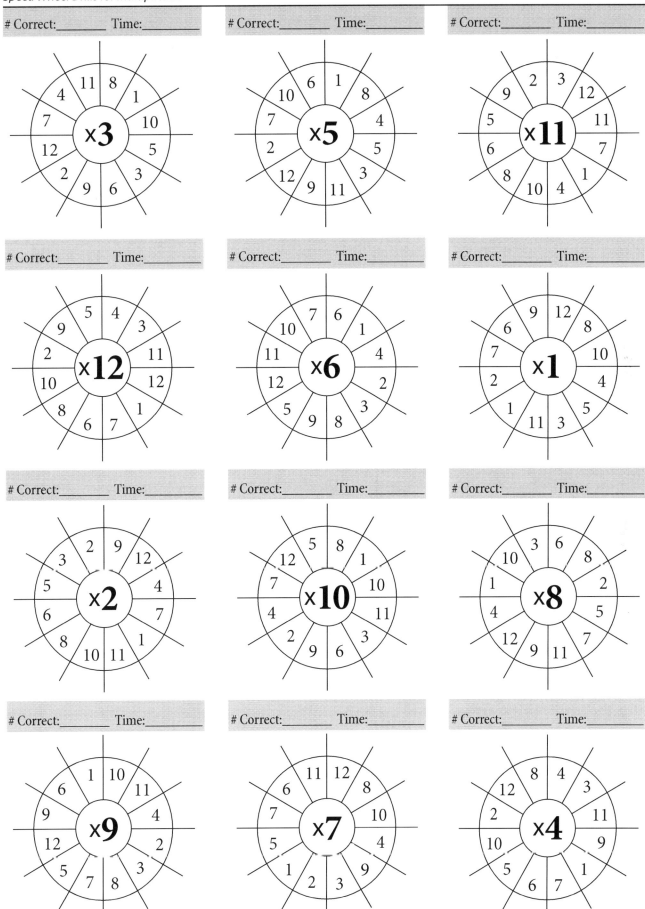

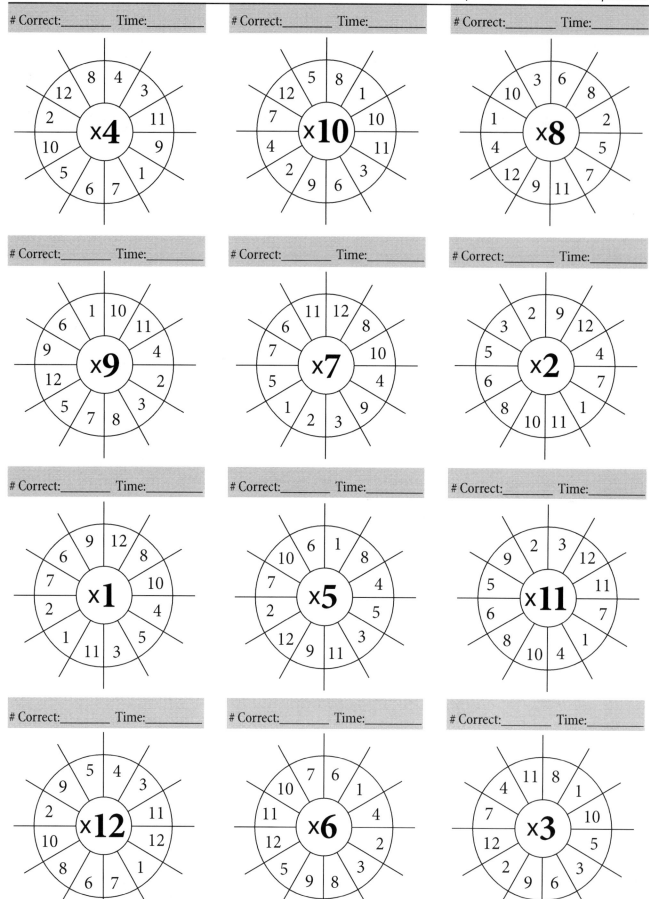

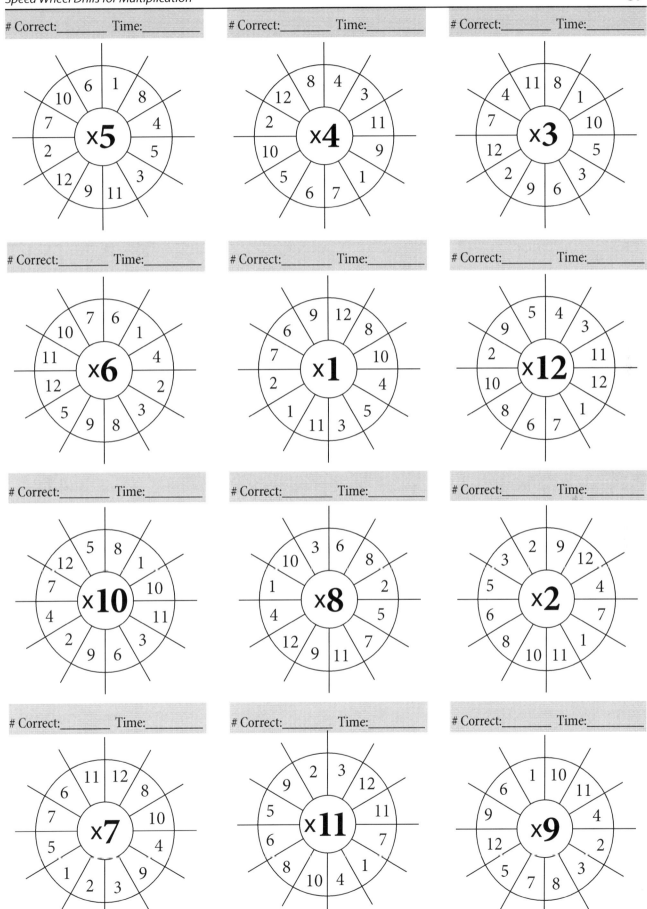

58 — Speed Wheel Drills for Multiplication

Correct:_____ Time:_____
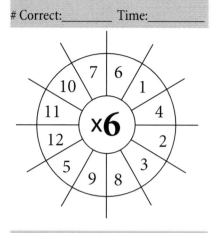

Correct:_____ Time:_____
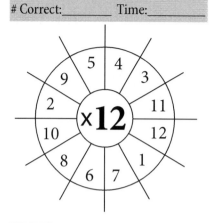

Correct:_____ Time:_____
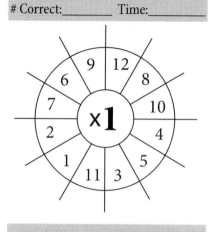

Correct:_____ Time:_____
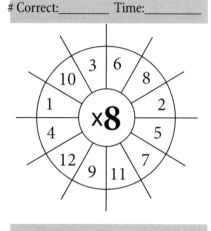

Correct:_____ Time:_____

Correct:_____ Time:_____
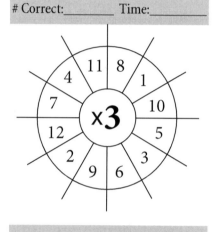

Correct:_____ Time:_____
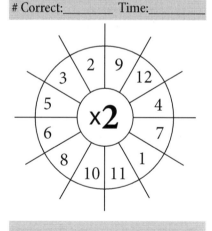

Correct:_____ Time:_____
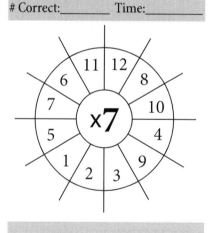

Correct:_____ Time:_____

Correct:_____ Time:_____
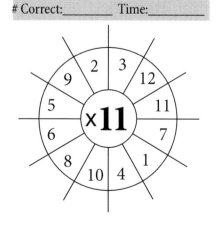

Correct:_____ Time:_____
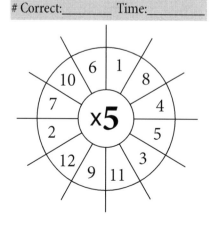

Correct:_____ Time:_____
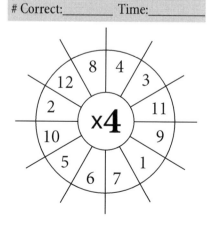

It is illegal to photocopy this page. Copyright © 2021, Richard W. Fisher

Speed Wheel Drills for Multiplication 59

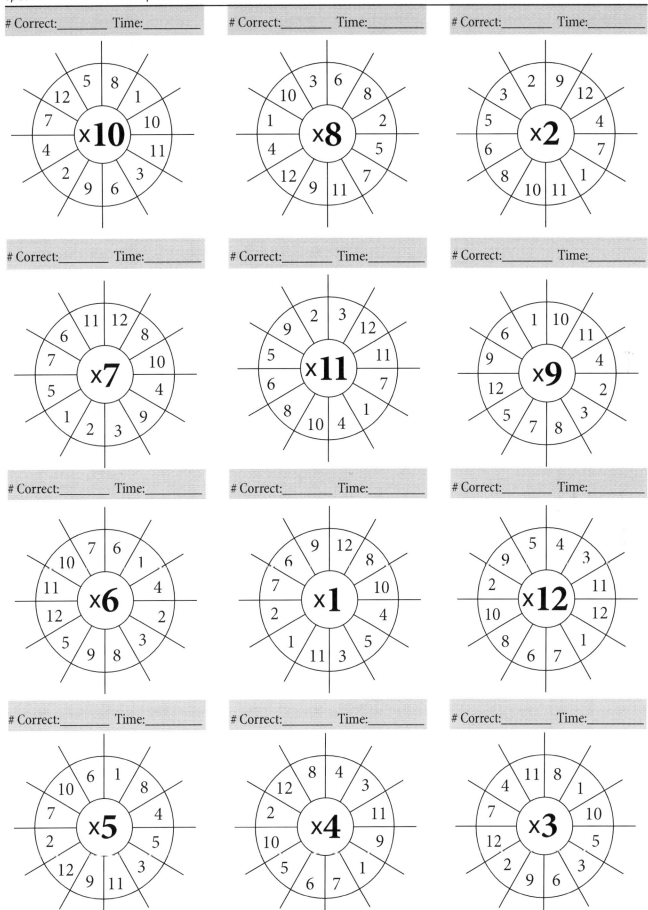

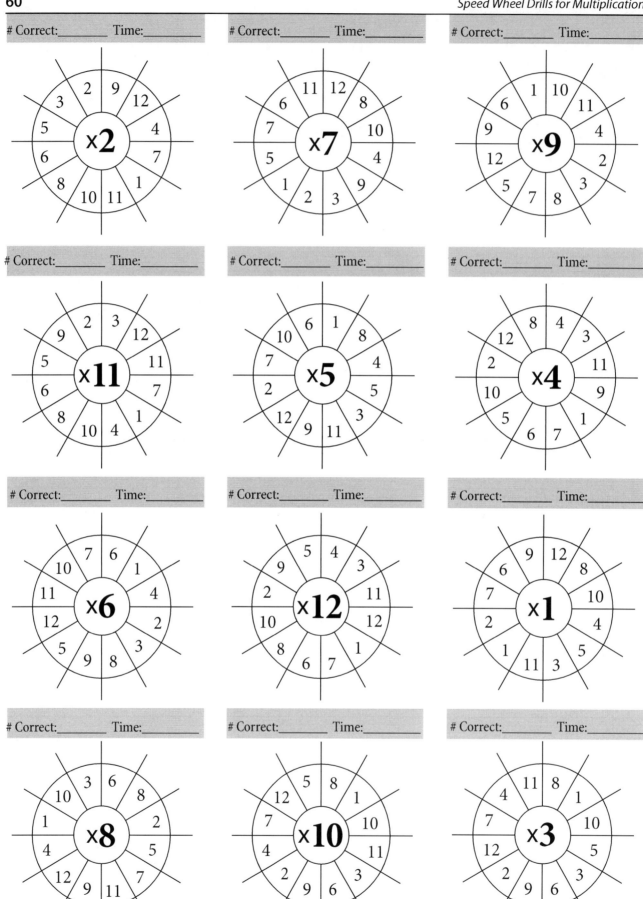

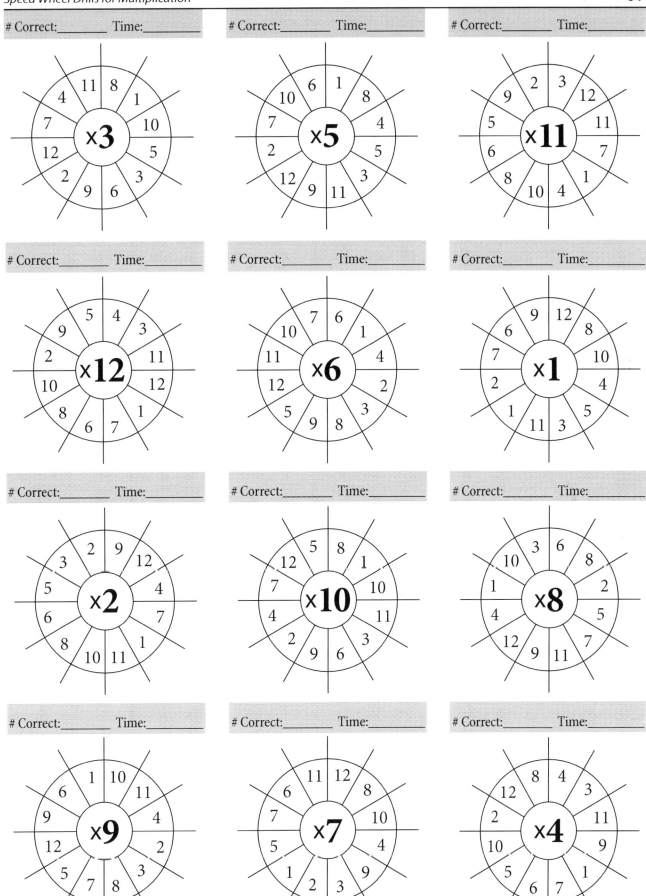

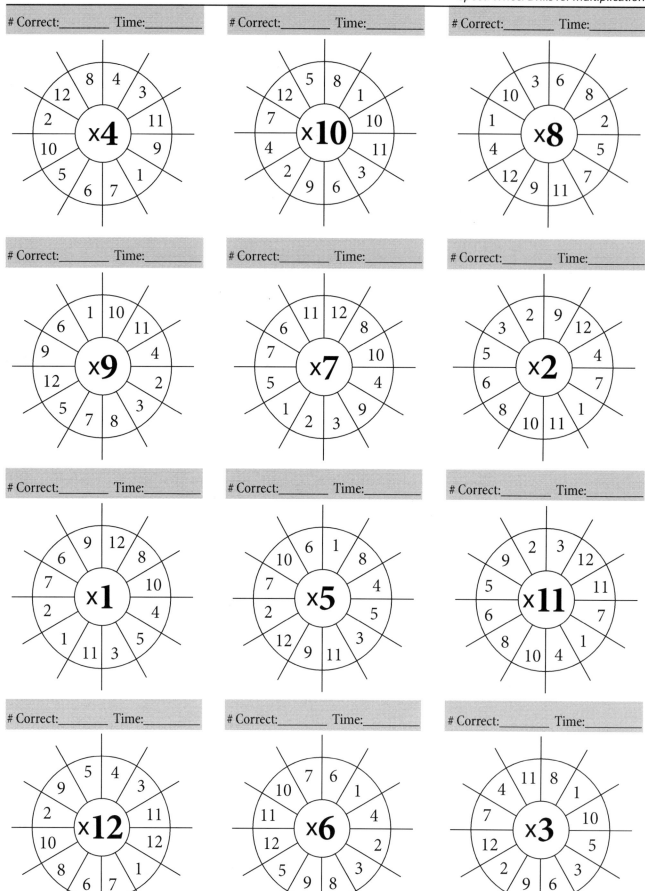

Speed Wheel Drills for Multiplication 63

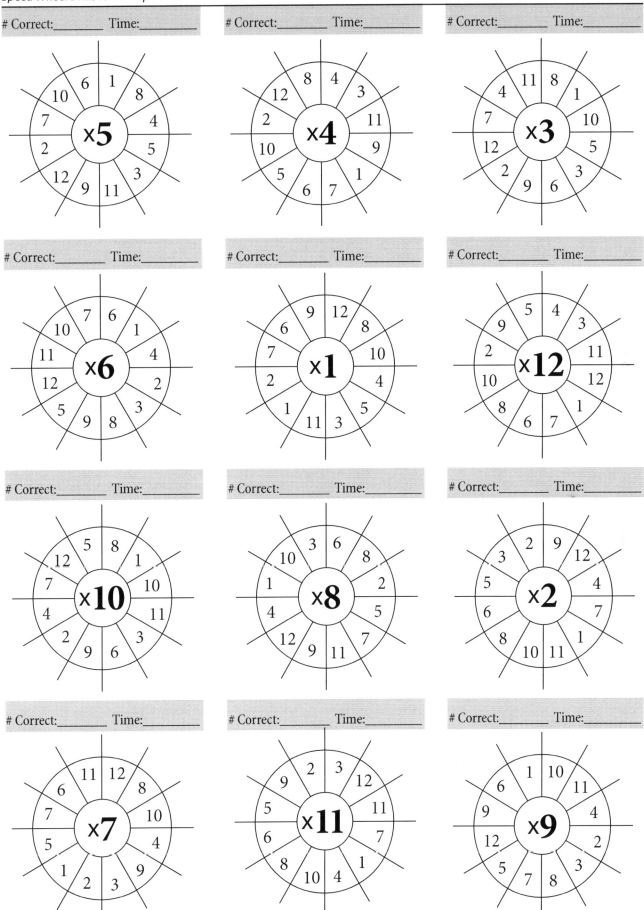

It is illegal to photocopy this page. Copyright © 2021, Richard W. Fisher

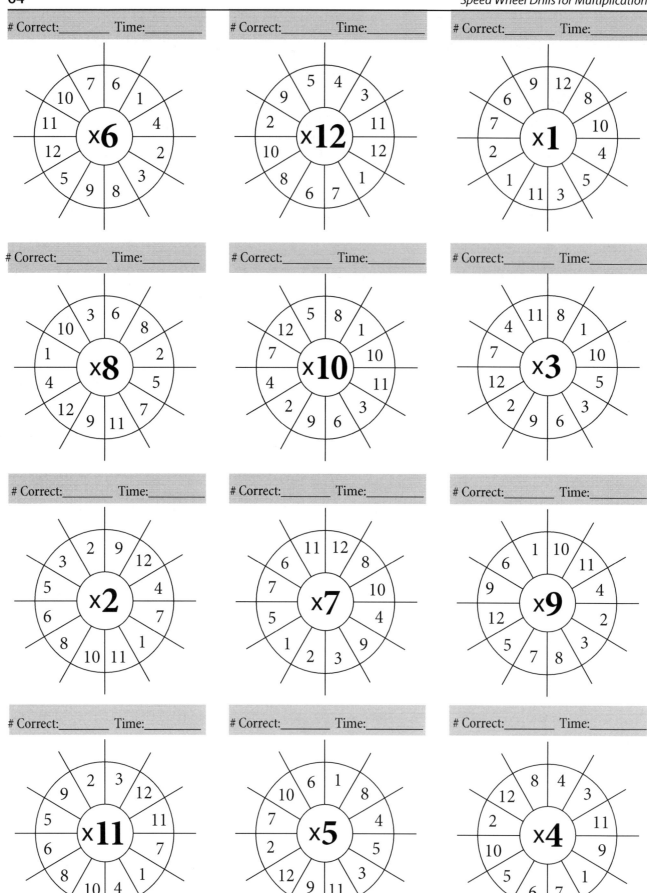

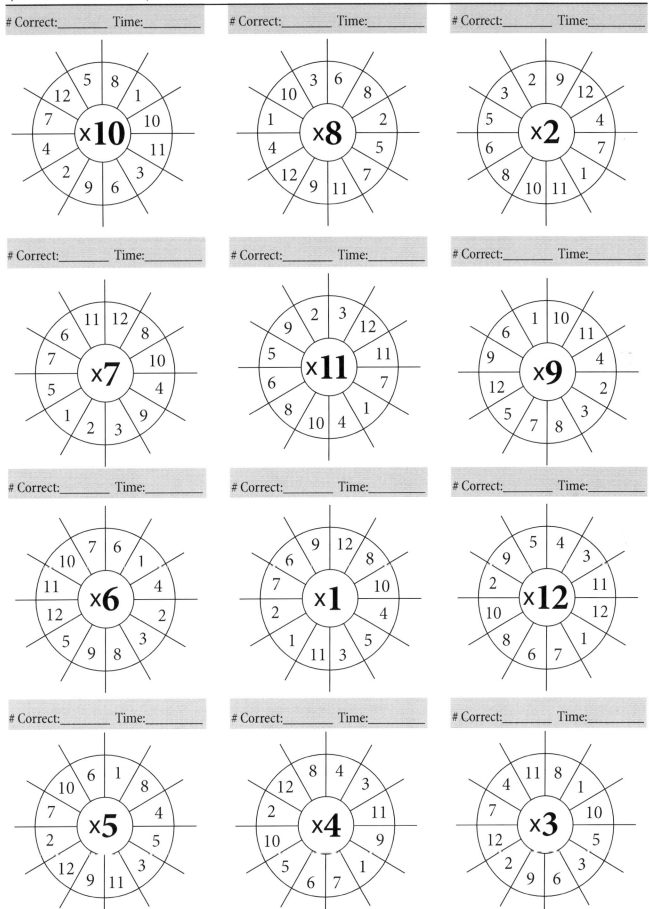

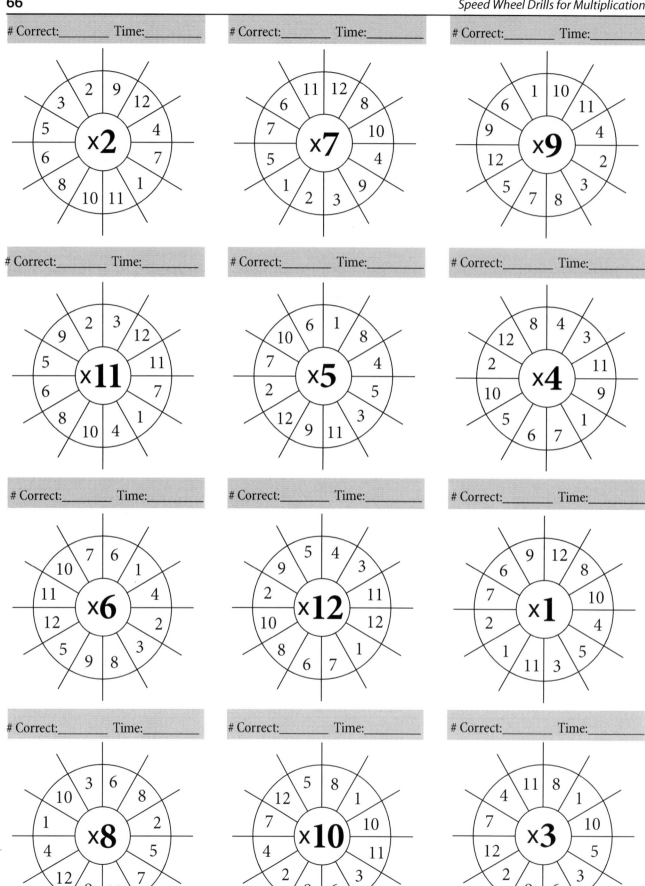

Speed Wheel Drills for Multiplication 67

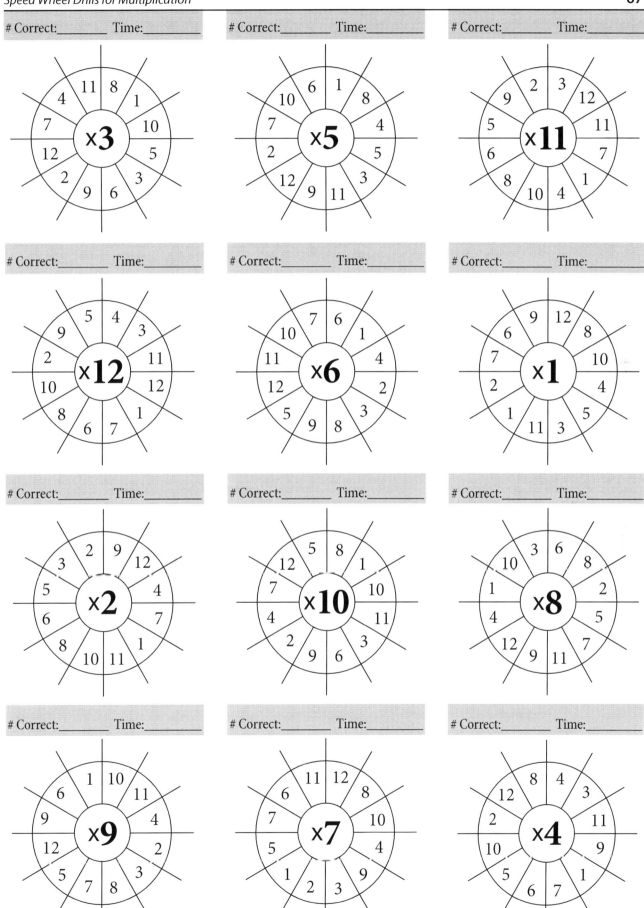

Speed Wheel Drills for Multiplication

Correct:_____ Time:_____

×4 wheel: 8, 4, 3, 11, 9, 1, 7, 6, 5, 10, 2, 12

Correct:_____ Time:_____

×10 wheel: 5, 8, 1, 10, 11, 3, 6, 9, 2, 4, 7, 12

Correct:_____ Time:_____

×8 wheel: 3, 6, 8, 2, 5, 7, 11, 9, 12, 4, 1, 10

Correct:_____ Time:_____

×9 wheel: 1, 10, 11, 4, 2, 3, 8, 7, 5, 12, 9, 6

Correct:_____ Time:_____

×7 wheel: 11, 12, 8, 10, 4, 9, 3, 2, 1, 5, 7, 6

Correct:_____ Time:_____

×2 wheel: 2, 9, 12, 4, 7, 1, 11, 10, 8, 6, 5, 3

Correct:_____ Time:_____

×1 wheel: 9, 12, 8, 10, 4, 5, 3, 11, 1, 2, 7, 6

Correct:_____ Time:_____

×5 wheel: 6, 1, 8, 4, 5, 3, 11, 9, 12, 2, 7, 10

Correct:_____ Time:_____

×11 wheel: 2, 3, 12, 11, 7, 1, 4, 10, 8, 6, 5, 9

Correct:_____ Time:_____

×12 wheel: 5, 4, 3, 11, 12, 1, 7, 6, 8, 10, 2, 9

Correct:_____ Time:_____

×6 wheel: 7, 6, 1, 4, 2, 3, 8, 9, 5, 12, 11, 10

Correct:_____ Time:_____

×3 wheel: 11, 8, 1, 10, 5, 3, 6, 9, 2, 12, 7, 4

Speed Wheel Drills for Multiplication

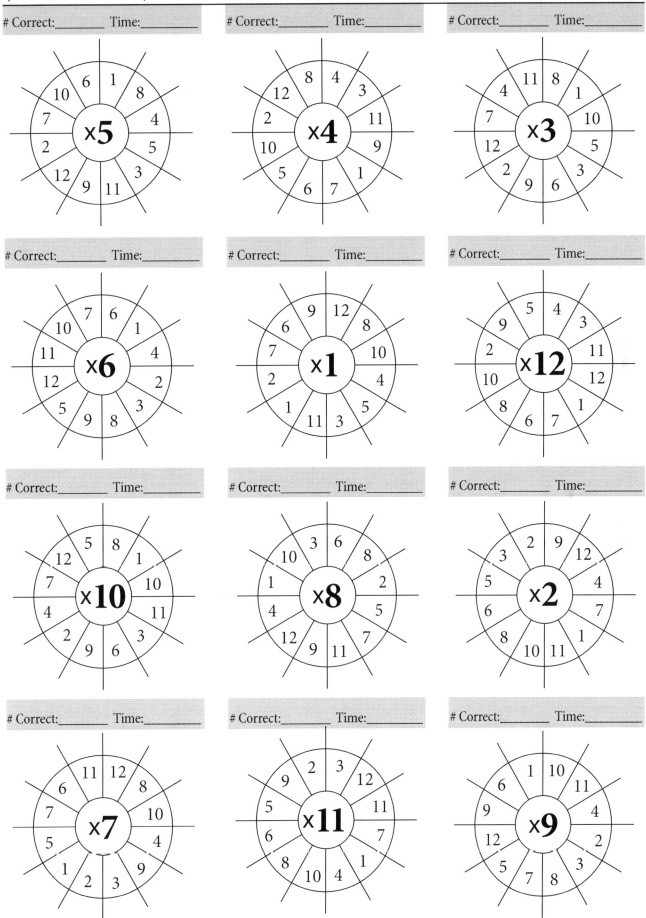

Speed Wheel Drills for Multiplication

Correct:_____ Time:_____
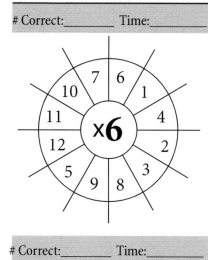

Correct:_____ Time:_____

Correct:_____ Time:_____
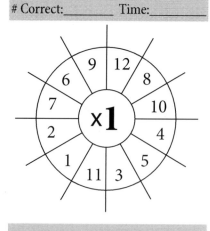

Correct:_____ Time:_____
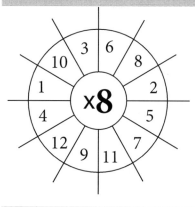

Correct:_____ Time:_____
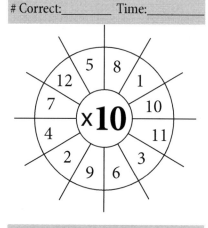

Correct:_____ Time:_____
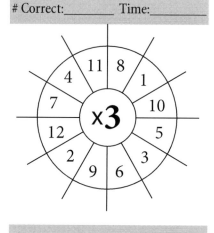

Correct:_____ Time:_____
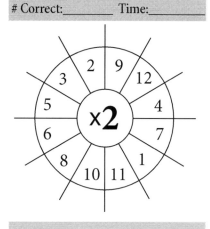

Correct:_____ Time:_____
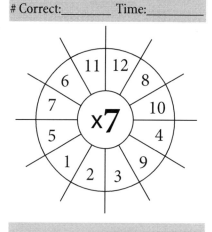

Correct:_____ Time:_____
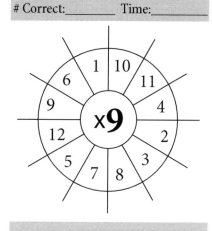

Correct:_____ Time:_____
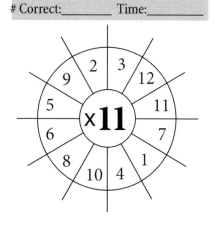

Correct:_____ Time:_____
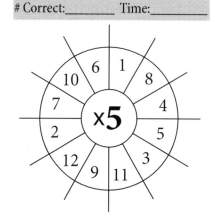

Correct:_____ Time:_____
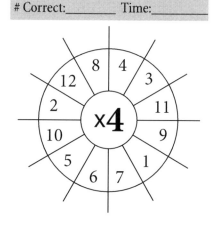

It is illegal to photocopy this page. Copyright © 2021, Richard W. Fisher

Speed Wheel Drills for Multiplication

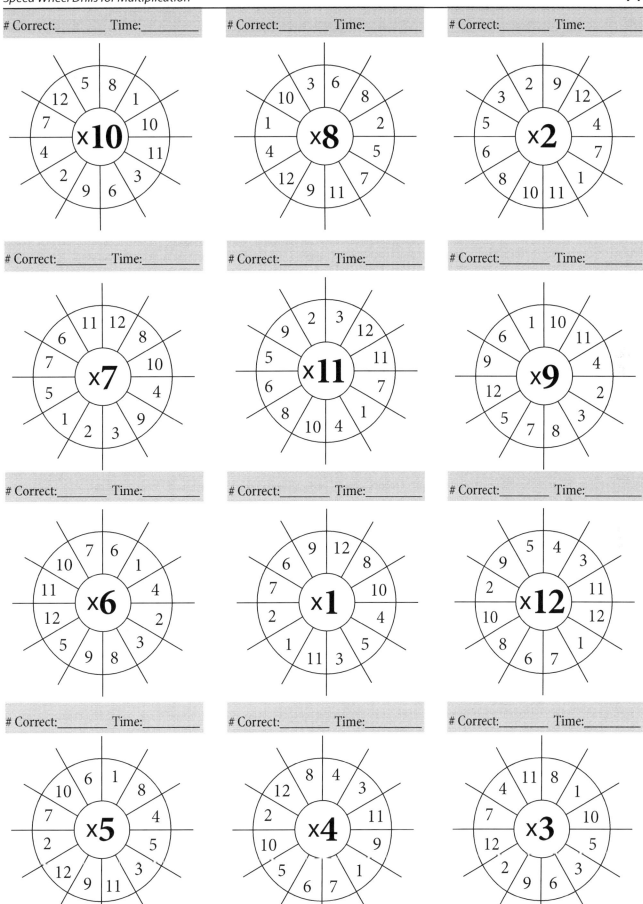

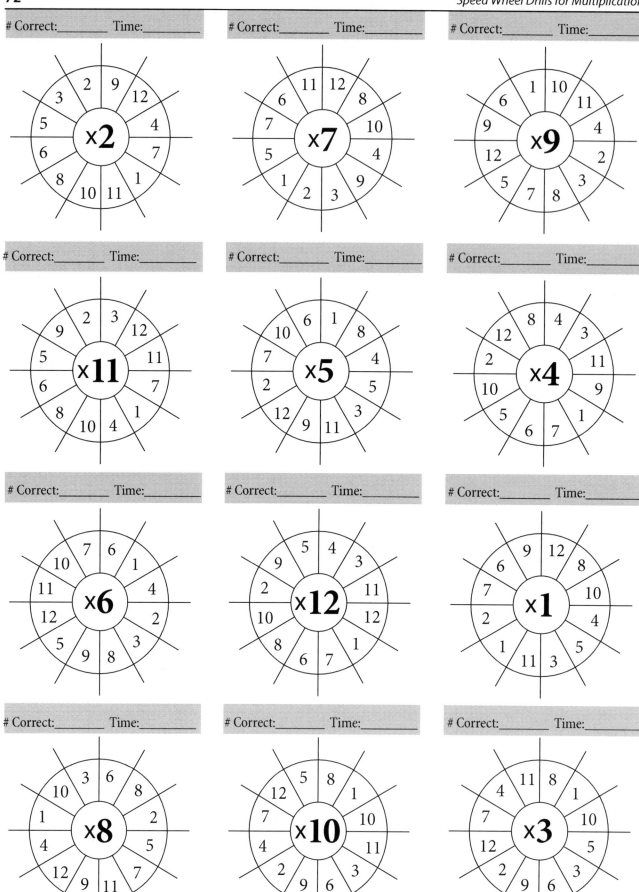

Speed Wheel Drills for Multiplication 73

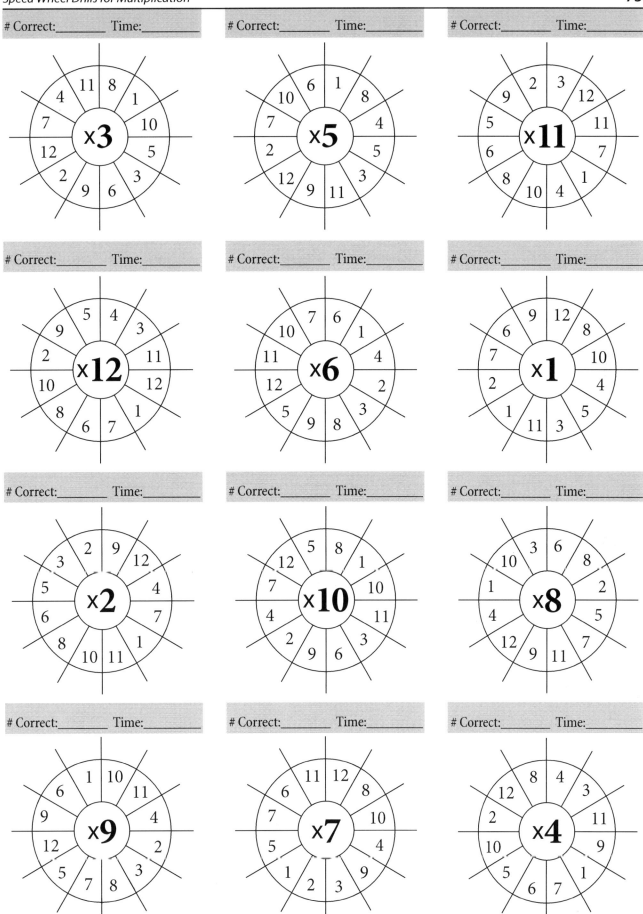

Correct:_____ Time:_____
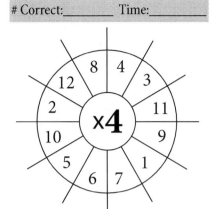

Correct:_____ Time:_____
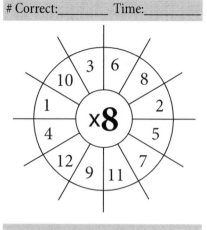

Correct:_____ Time:_____
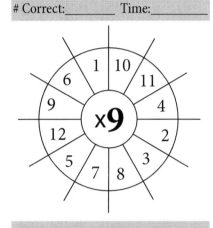

Correct:_____ Time:_____
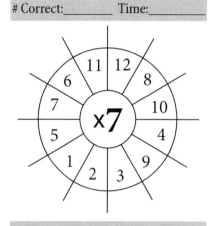

Correct:_____ Time:_____
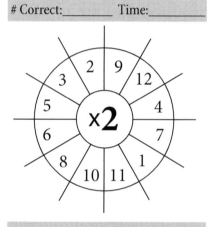

Correct:_____ Time:_____
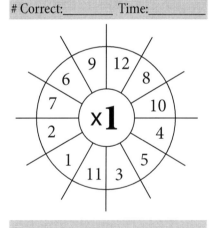

Correct:_____ Time:_____

Correct:_____ Time:_____

Correct:_____ Time:_____
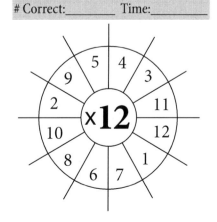

Correct:_____ Time:_____
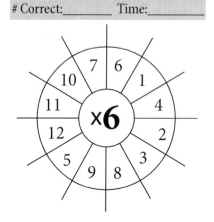

Correct:_____ Time:_____
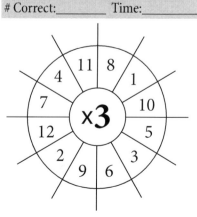

Speed Wheel Drills for Multiplication 75

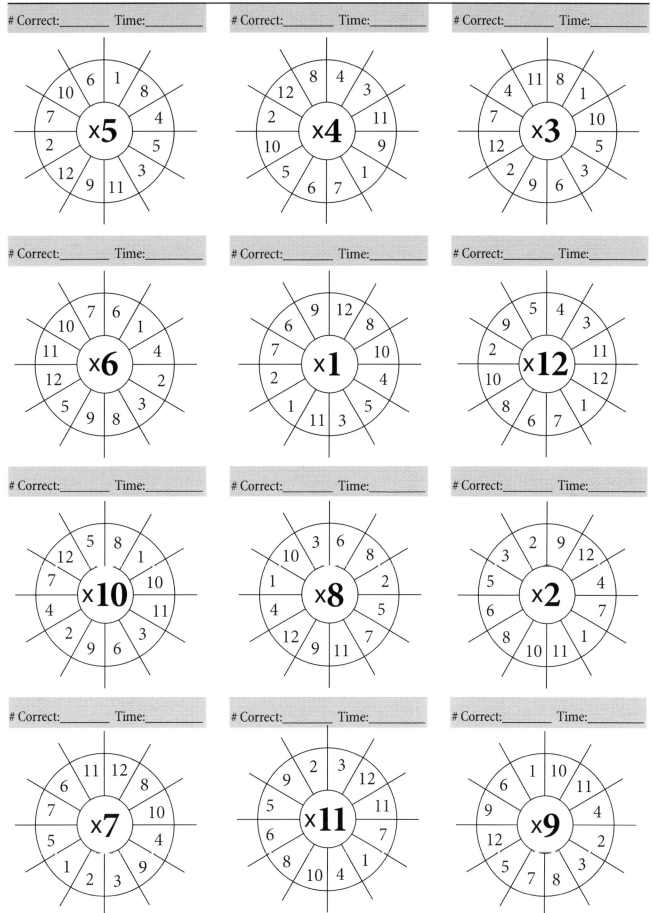

It is illegal to photocopy this page. Copyright © 2021, Richard W. Fisher

Correct:_____ Time:_____

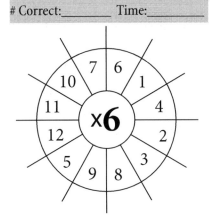

Correct:_____ Time:_____

Correct:_____ Time:_____

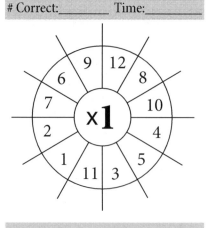

Correct:_____ Time:_____

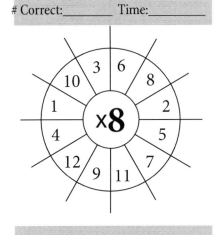

Correct:_____ Time:_____

Correct:_____ Time:_____

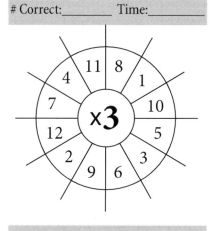

Correct:_____ Time:_____

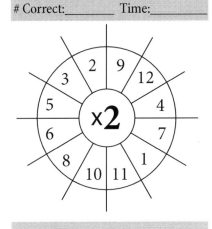

Correct:_____ Time:_____

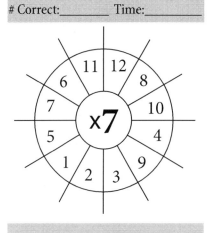

Correct:_____ Time:_____

Correct:_____ Time:_____

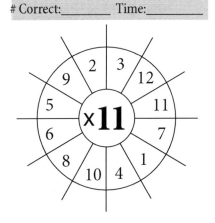

Correct:_____ Time:_____

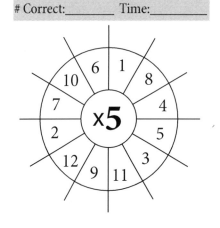

Correct:_____ Time:_____

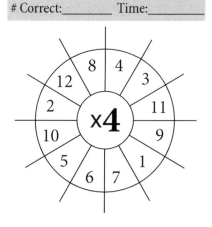

Speed Wheel Drills for Multiplication

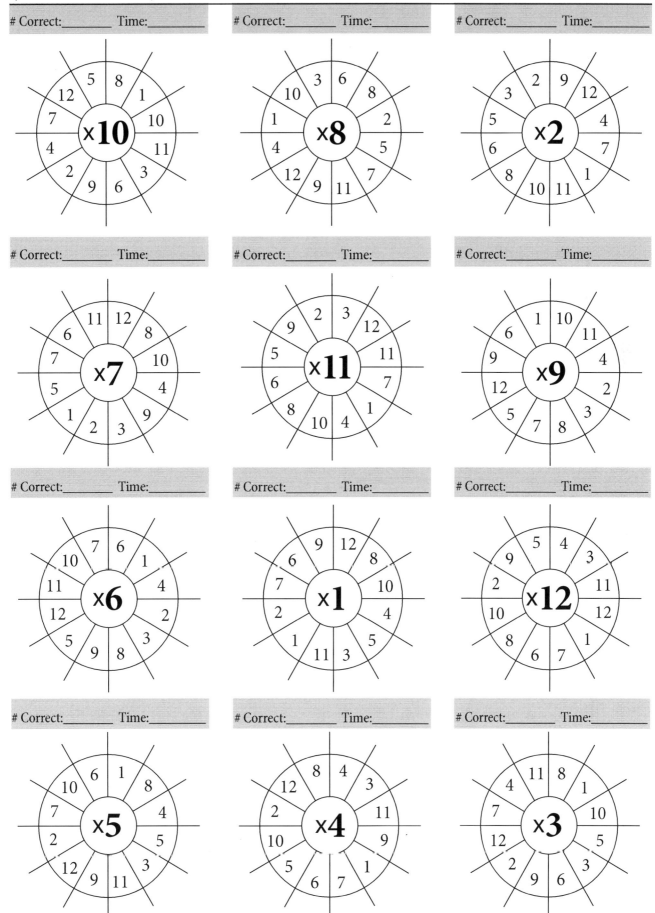

Correct:_____ Time:_____

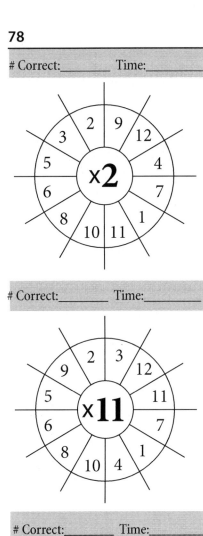

Correct:_____ Time:_____

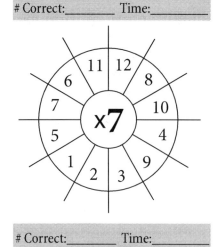

Correct:_____ Time:_____

Correct:_____ Time:_____

Correct:_____ Time:_____

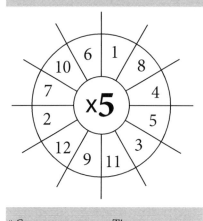

Correct:_____ Time:_____

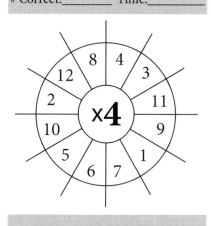

Correct:_____ Time:_____

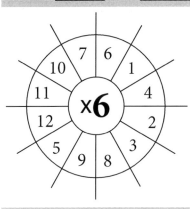

Correct:_____ Time:_____

Correct:_____ Time:_____

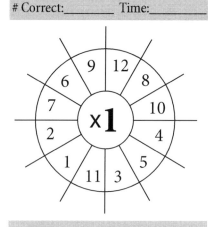

Correct:_____ Time:_____

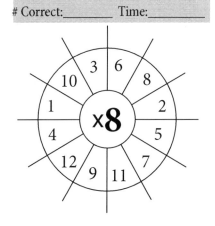

Correct:_____ Time:_____

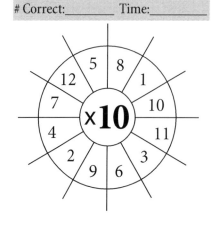

Correct:_____ Time:_____

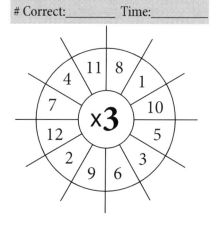

Speed Wheel Drills for Multiplication

Correct:_____ Time:_____

Correct:_____ Time:_____

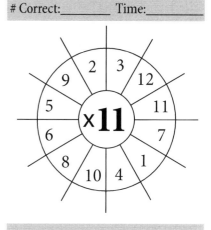

Correct:_____ Time:_____

Correct:_____ Time:_____

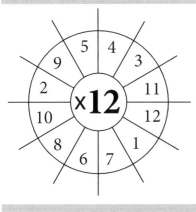

Correct:_____ Time:_____

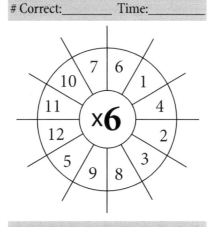

Correct:_____ Time:_____

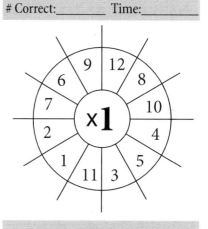

Correct:_____ Time:_____

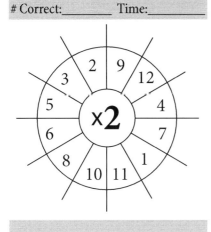

Correct:_____ Time:_____

Correct:_____ Time:_____

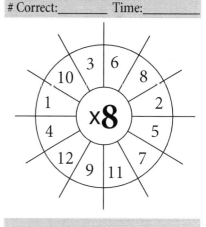

Correct:_____ Time:_____

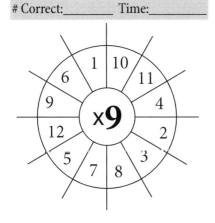

Correct:_____ Time:_____

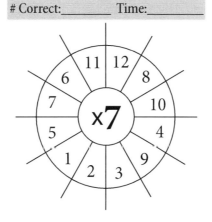

Correct:_____ Time:_____

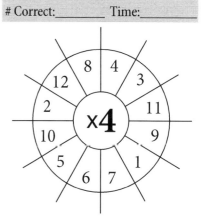

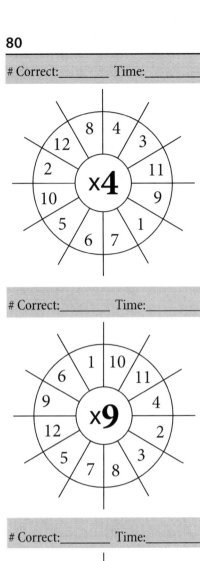
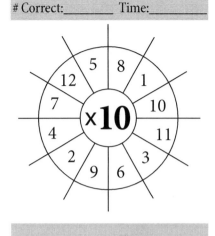
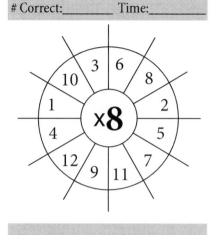
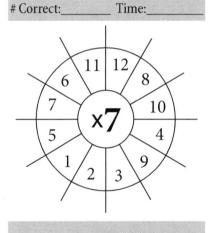
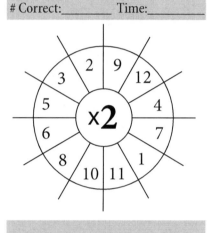
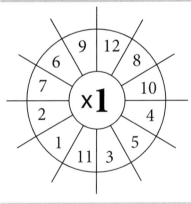
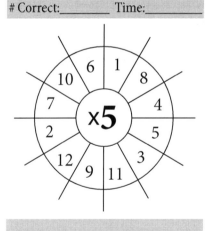
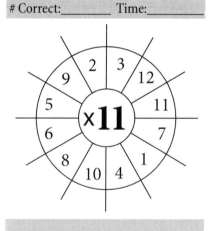
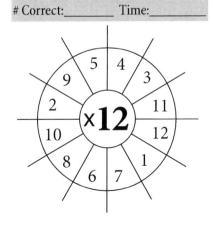
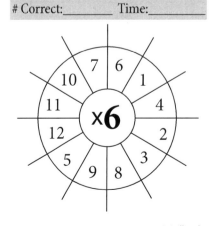
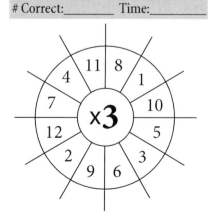

Speed Wheel Drills for Multiplication 81

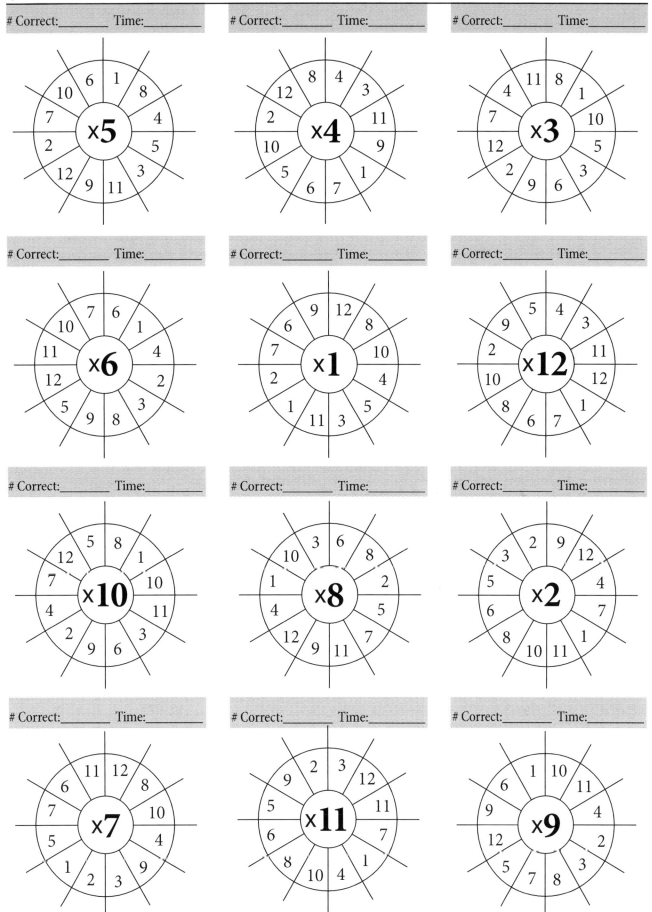

It is illegal to photocopy this page. Copyright © 2021, Richard W. Fisher

Speed Wheel Drills for Multiplication

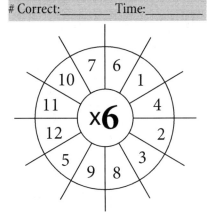
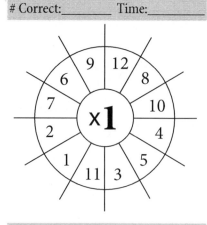
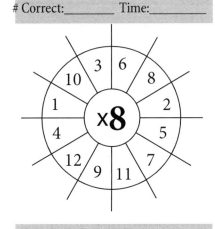
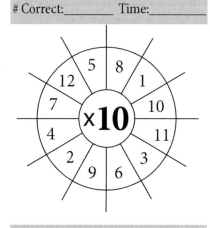
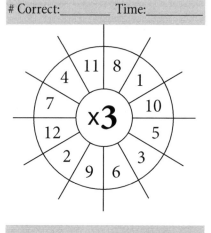
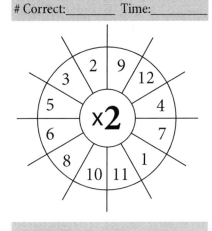
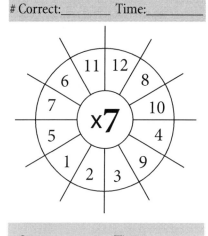
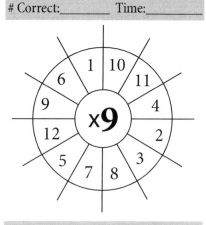
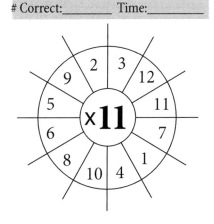
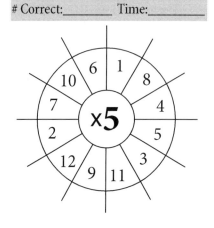
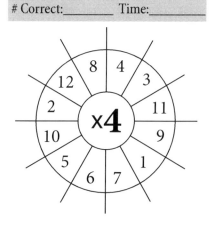

It is illegal to photocopy this page. Copyright © 2021, Richard W. Fisher

Speed Wheel Drills for Multiplication 83

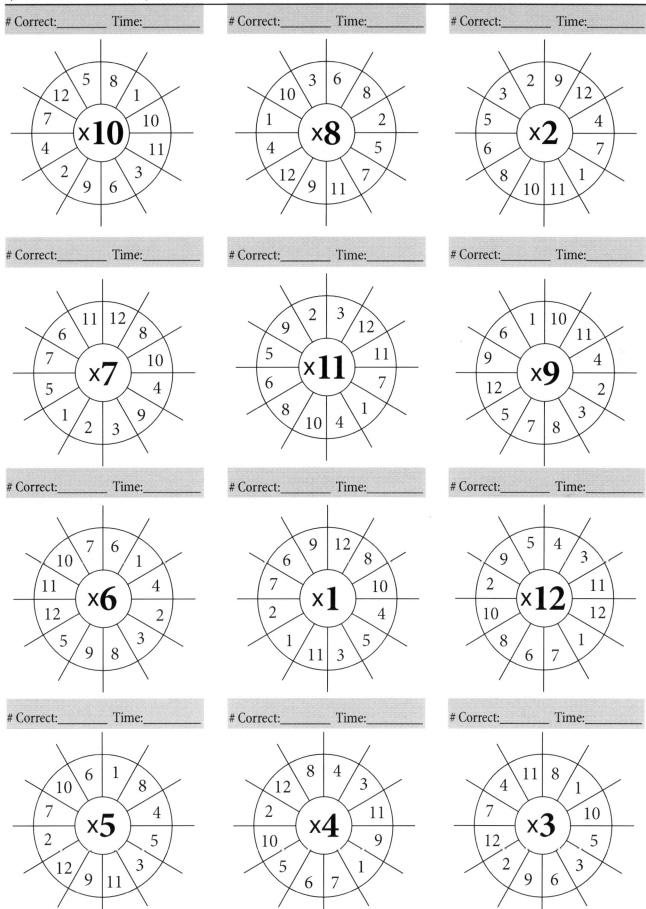

It is illegal to photocopy this page. Copyright © 2021, Richard W. Fisher

84 Speed Wheel Drills for Multiplication

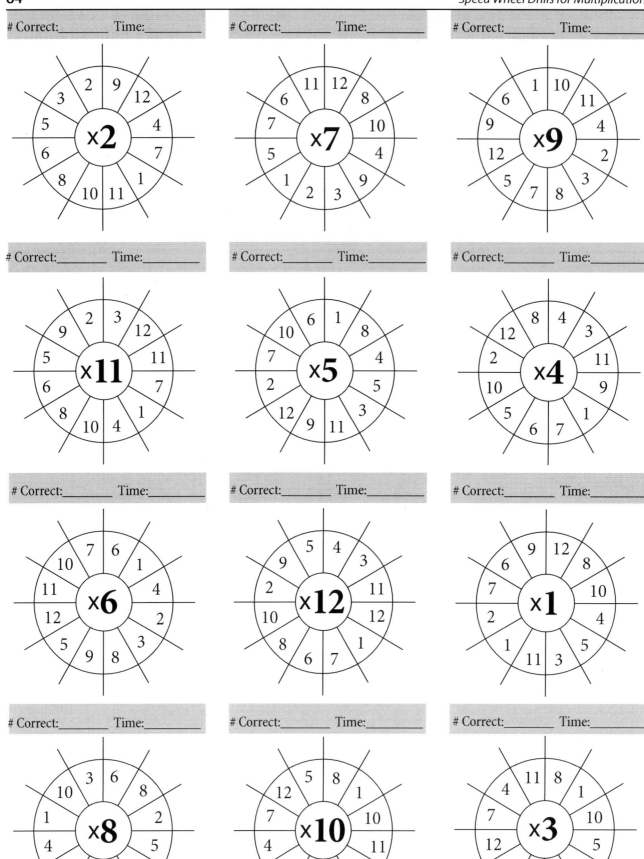

It is illegal to photocopy this page. Copyright © 2021, Richard W. Fisher

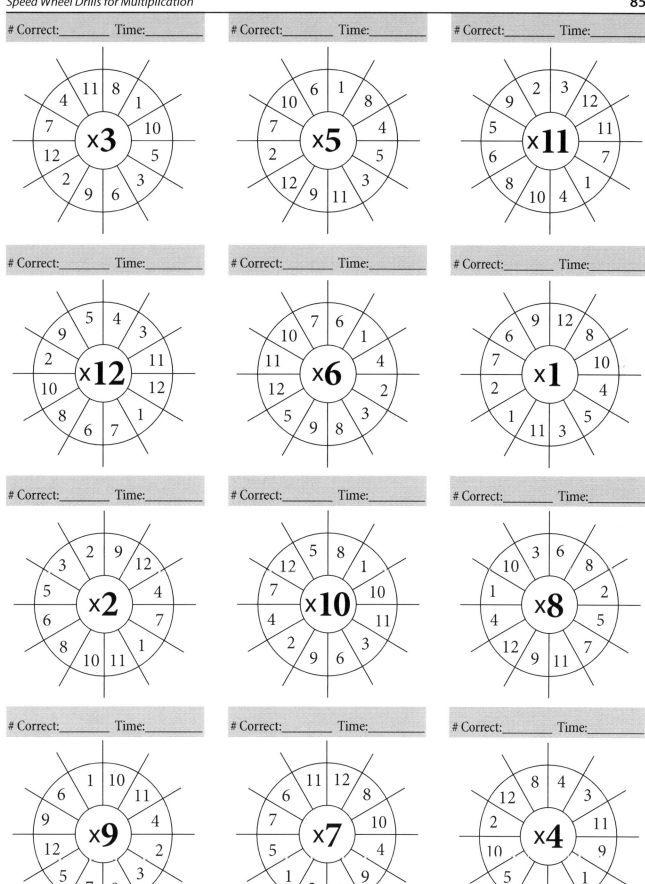

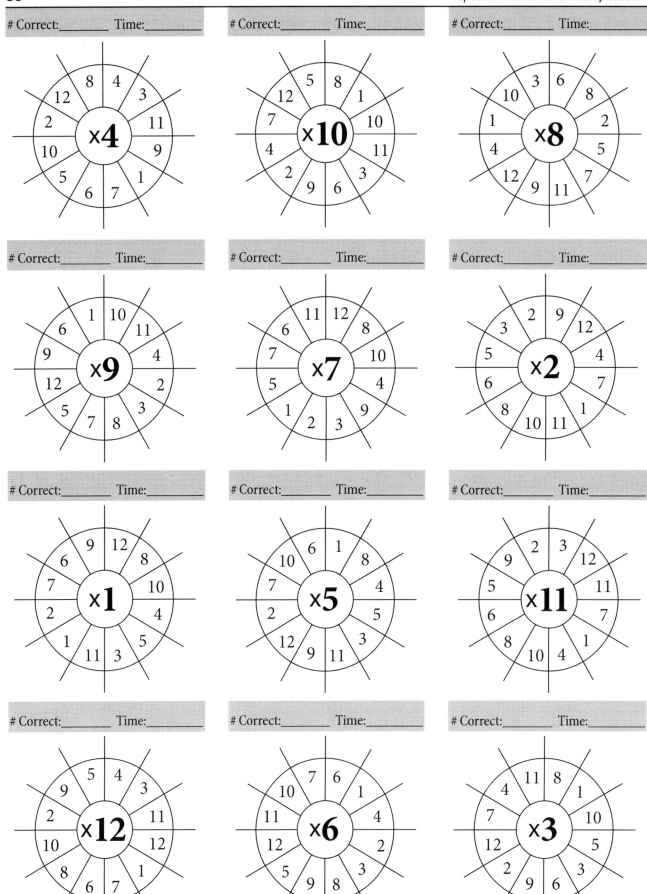

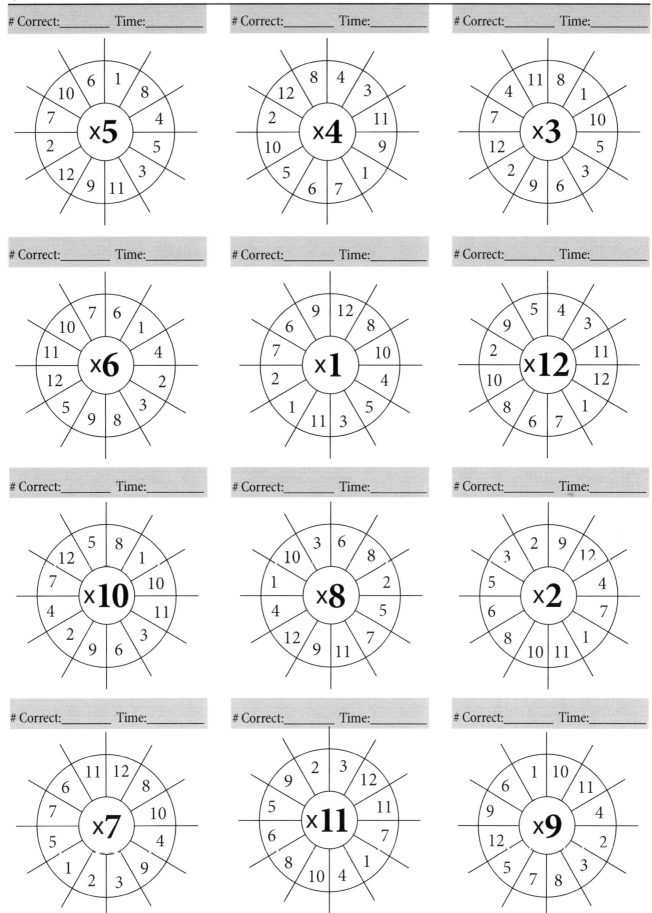

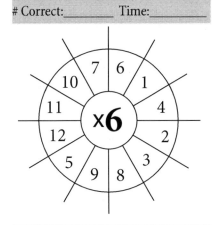
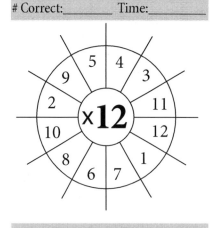
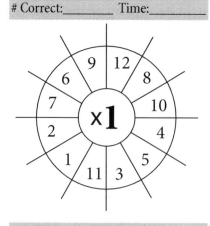
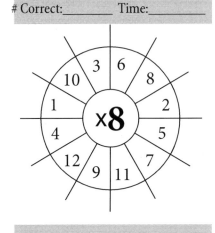
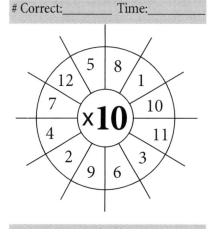
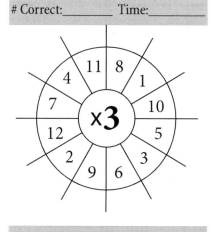
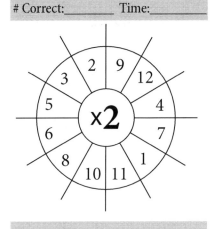
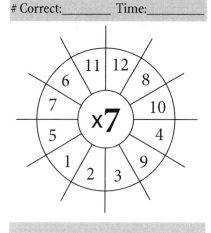

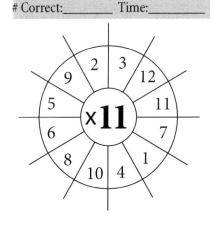
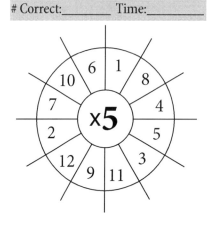
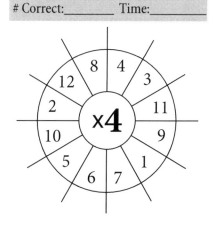

Speed Wheel Drills for Multiplication

Correct:_____ Time:_____

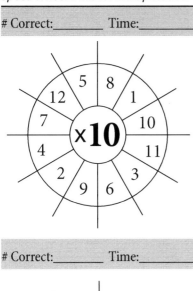

Correct:_____ Time:_____

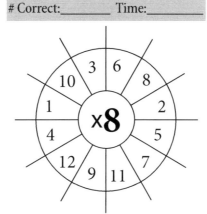

Correct:_____ Time:_____

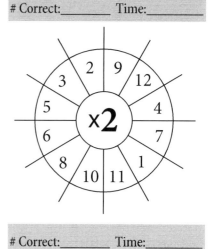

Correct:_____ Time:_____

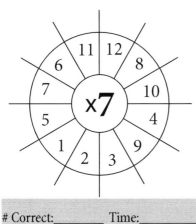

Correct:_____ Time:_____

Correct:_____ Time:_____

Correct:_____ Time:_____

Correct:_____ Time:_____

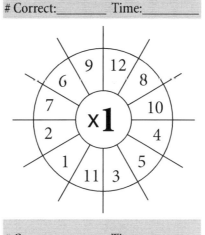

Correct:_____ Time:_____

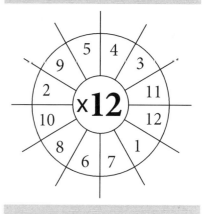

Correct:_____ Time:_____

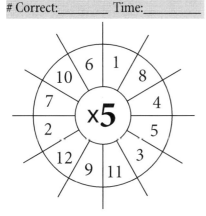

Correct:_____ Time:_____

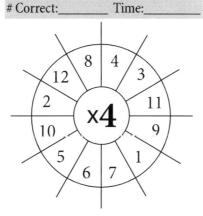

Correct:_____ Time:_____

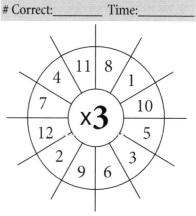

It is illegal to photocopy this page. Copyright © 2021, Richard W. Fisher

Correct:_____ Time:_____   # Correct:_____ Time:_____   # Correct:_____ Time:_____

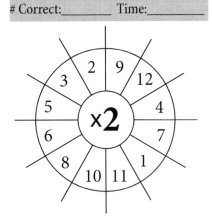 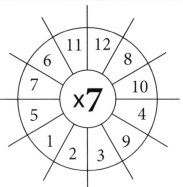 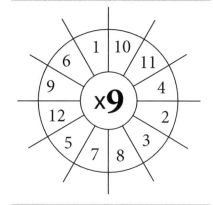

Correct:_____ Time:_____   # Correct:_____ Time:_____   # Correct:_____ Time:_____

 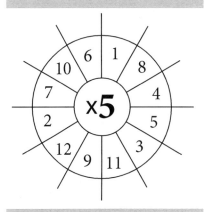 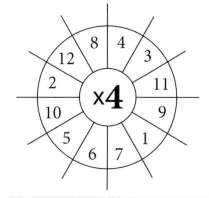

Correct:_____ Time:_____   # Correct:_____ Time:_____   # Correct:_____ Time:_____

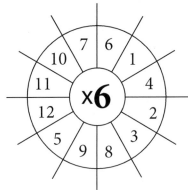 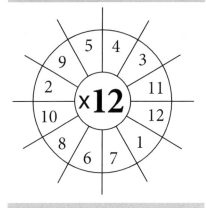 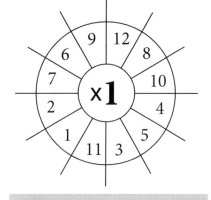

Correct:_____ Time:_____   # Correct:_____ Time:_____   # Correct:_____ Time:_____

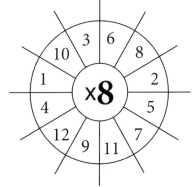 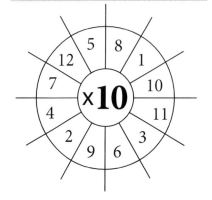 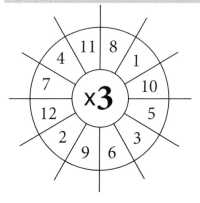

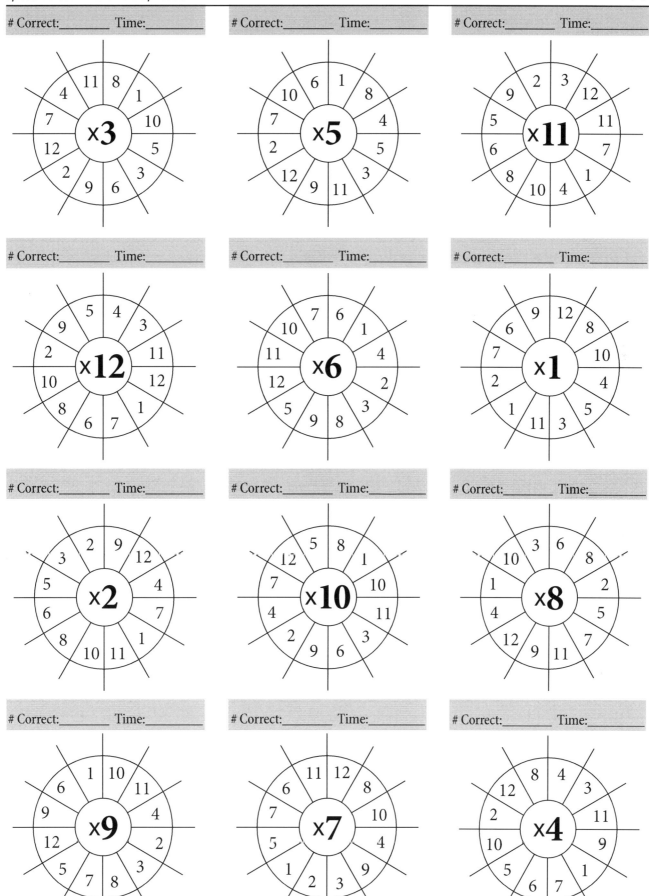

Speed Wheel Drills for Multiplication

Correct:_____ Time:_____
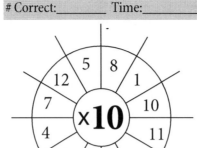

Correct:_____ Time:_____
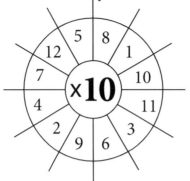

Correct:_____ Time:_____
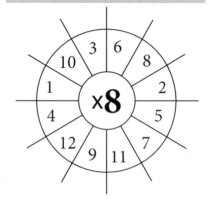

Correct:_____ Time:_____

Correct:_____ Time:_____
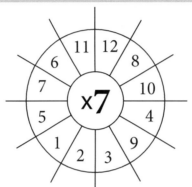

Correct:_____ Time:_____
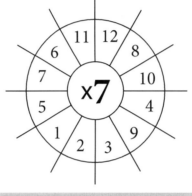

Correct:_____ Time:_____
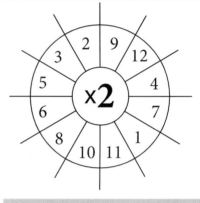

Correct:_____ Time:_____
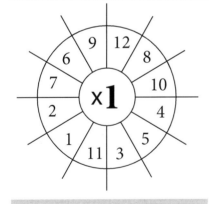

Correct:_____ Time:_____
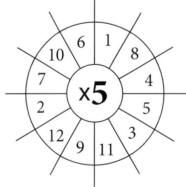

Correct:_____ Time:_____
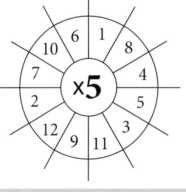

Correct:_____ Time:_____

Correct:_____ Time:_____
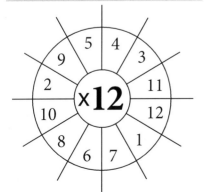

Correct:_____ Time:_____
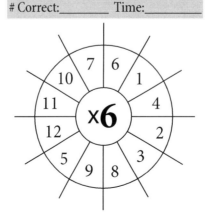

It is illegal to photocopy this page. Copyright © 2021, Richard W. Fisher

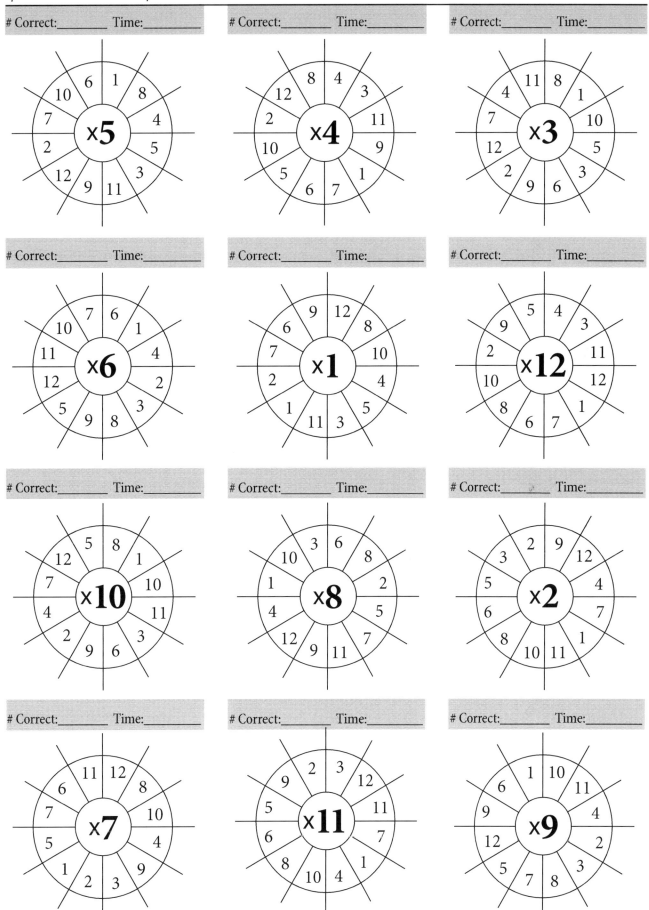

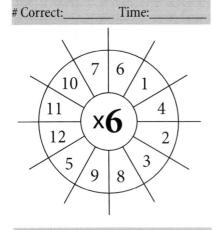
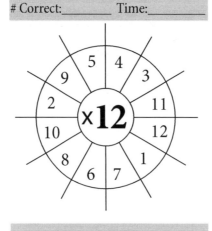
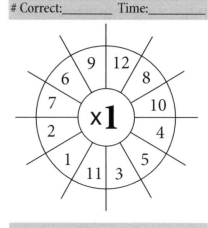
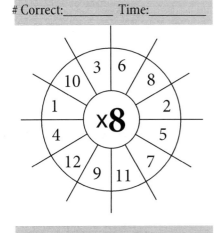
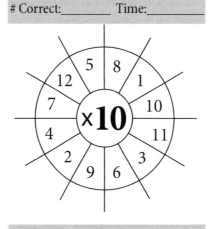
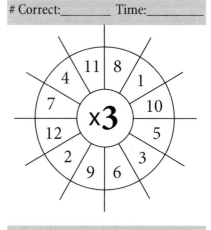
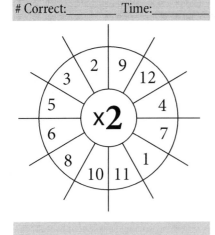
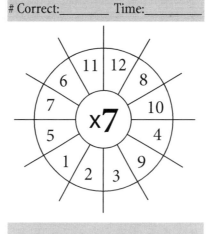

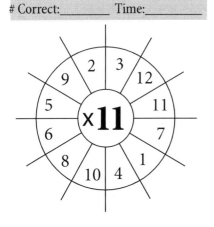
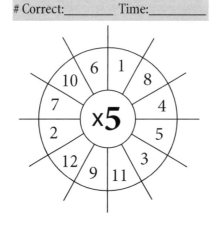
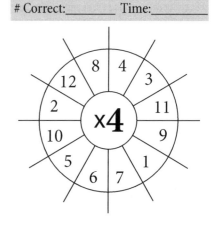

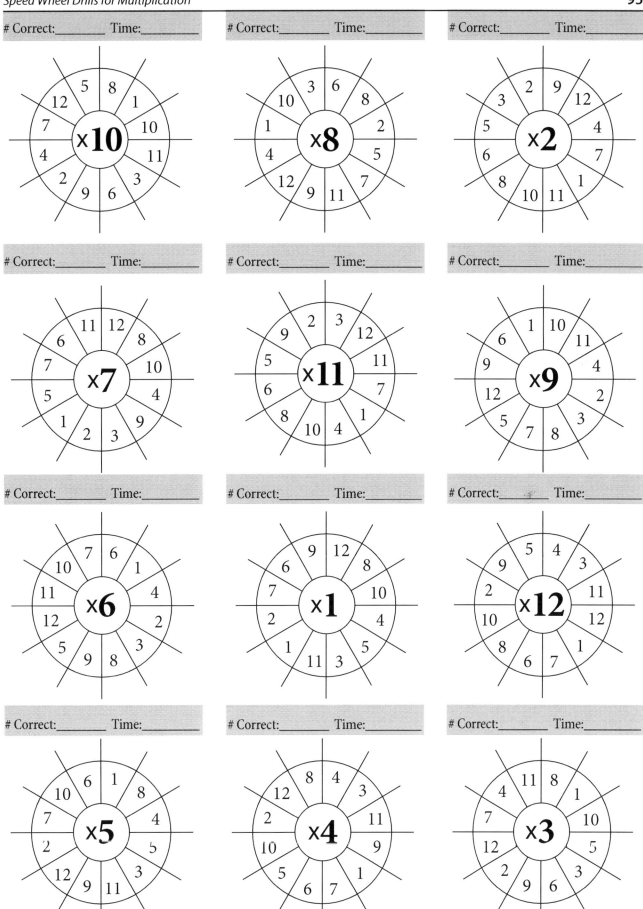

Speed Wheel Drills for Multiplication

Correct:_____ Time:_____
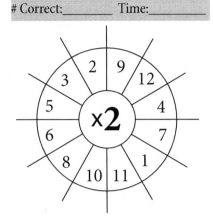

Correct:_____ Time:_____
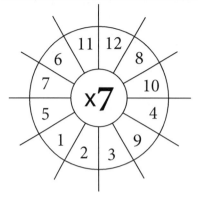

Correct:_____ Time:_____
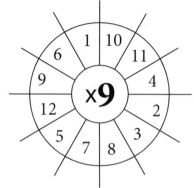

Correct:_____ Time:_____

Correct:_____ Time:_____
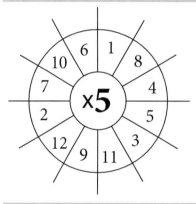

Correct:_____ Time:_____
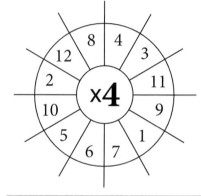

Correct:_____ Time:_____
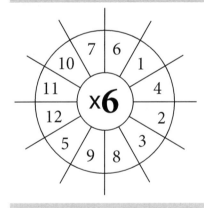

Correct:_____ Time:_____
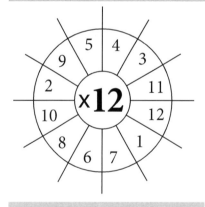

Correct:_____ Time:_____
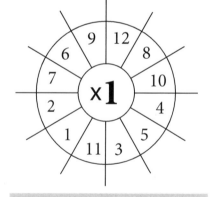

Correct:_____ Time:_____
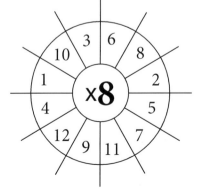

Correct:_____ Time:_____
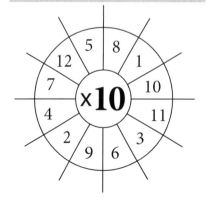

Correct:_____ Time:_____
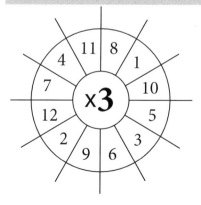

It is illegal to photocopy this page. Copyright © 2021, Richard W. Fisher

Speed Wheel Drills for Multiplication

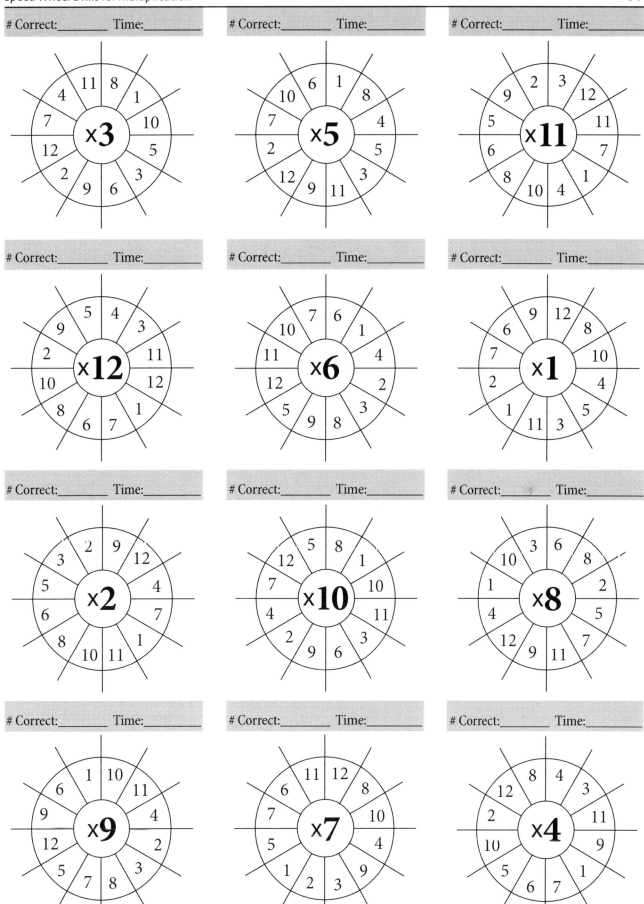

It is illegal to photocopy this page. Copyright © 2021, Richard W. Fisher

Speed Wheel Drills for Multiplication

# Correct:_____ Time:_____	# Correct:_____ Time:_____	# Correct:_____ Time:_____

  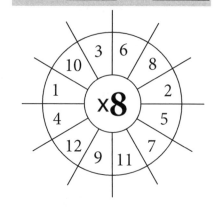

# Correct:_____ Time:_____	# Correct:_____ Time:_____	# Correct:_____ Time:_____

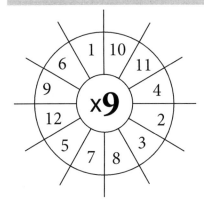  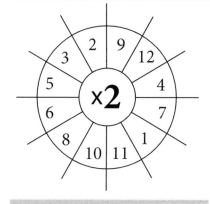

# Correct:_____ Time:_____	# Correct:_____ Time:_____	# Correct:_____ Time:_____

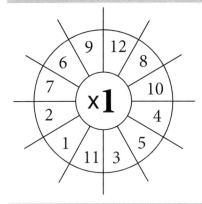 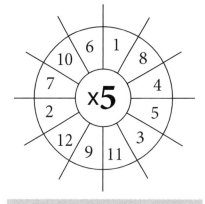 

# Correct:_____ Time:_____	# Correct:_____ Time:_____	# Correct:_____ Time:_____

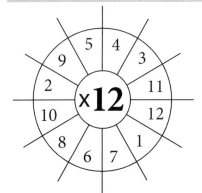 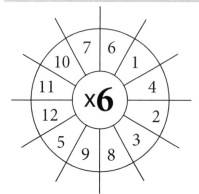 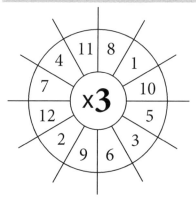

It is illegal to photocopy this page. Copyright © 2021, Richard W. Fisher

Speed Wheel Drills for Multiplication

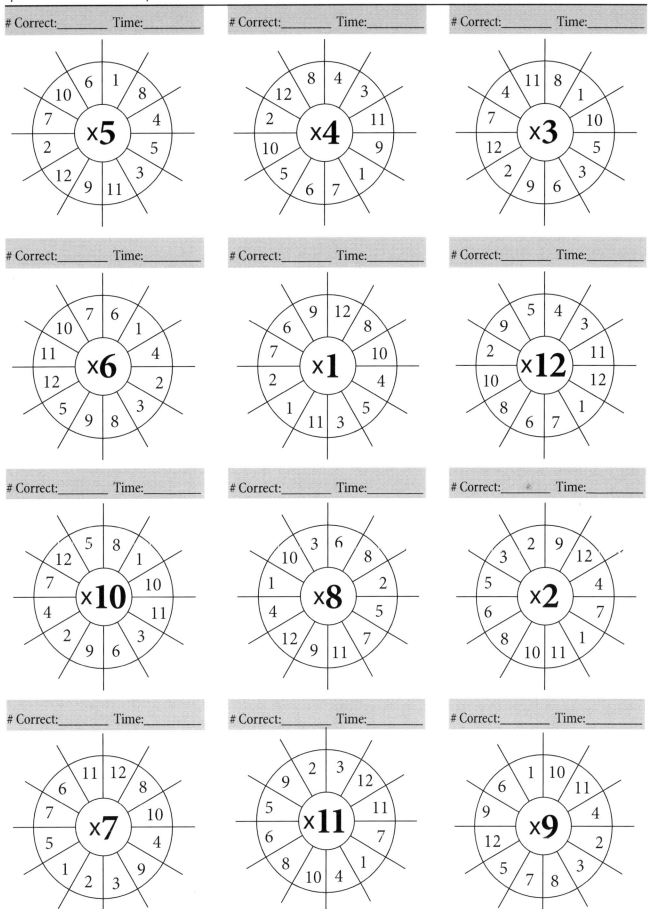

# Correct:_____ Time:_____	# Correct:_____ Time:_____	# Correct:_____ Time:_____
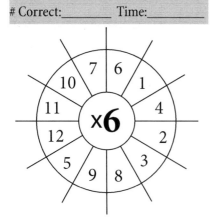	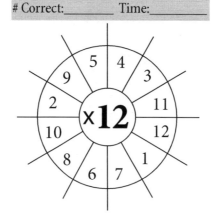	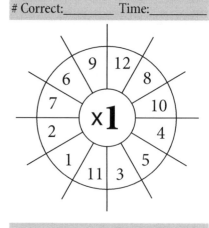
# Correct:_____ Time:_____	# Correct:_____ Time:_____	# Correct:_____ Time:_____
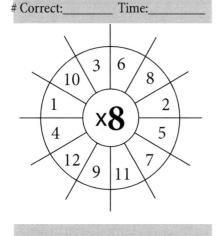	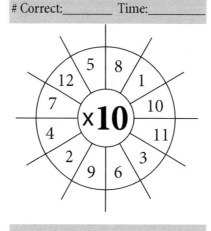	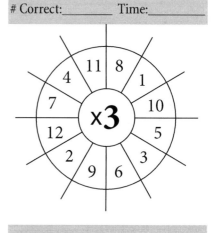
# Correct:_____ Time:_____	# Correct:_____ Time:_____	# Correct:_____ Time:_____
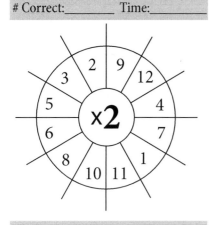	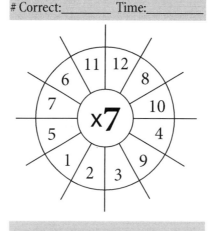	
# Correct:_____ Time:_____	# Correct:_____ Time:_____	# Correct:_____ Time:_____
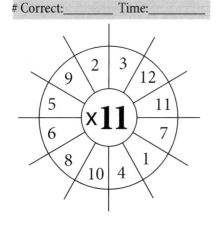	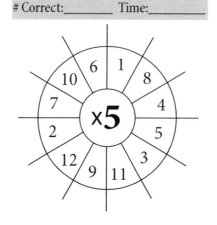	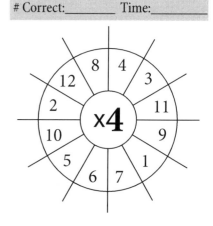

Speed Wheel Drills for Multiplication **101**

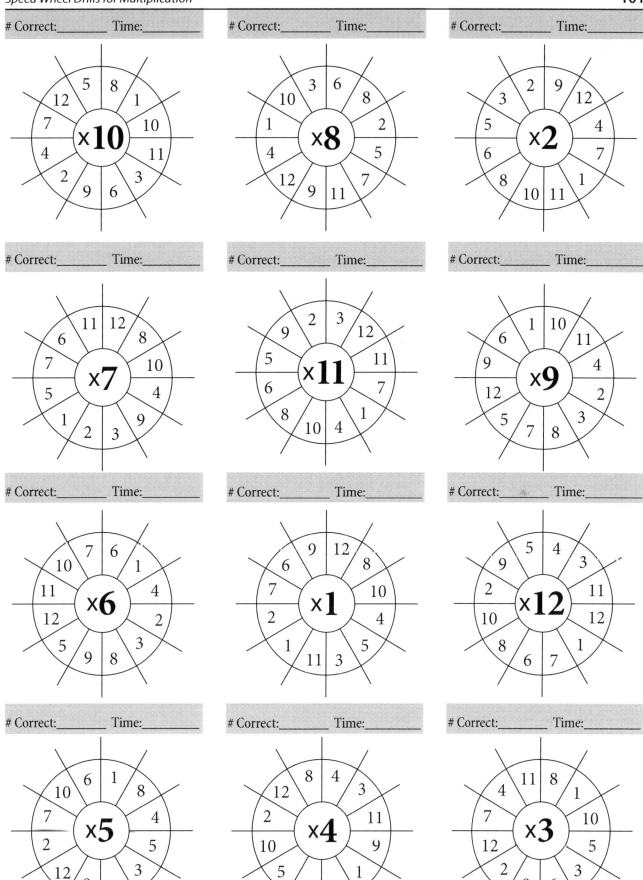

Correct:_____ Time:_____ # Correct:_____ Time:_____ # Correct:_____ Time:_____

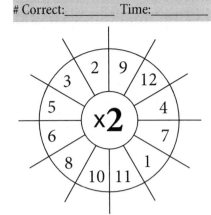 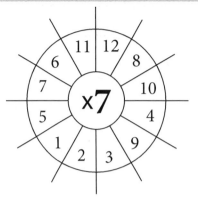 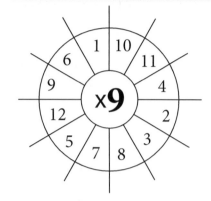

Correct:_____ Time:_____ # Correct:_____ Time:_____ # Correct:_____ Time:_____

 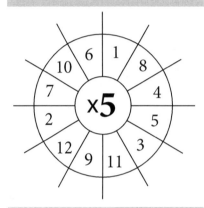 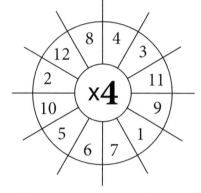

Correct:_____ Time:_____ # Correct:_____ Time:_____ # Correct:_____ Time:_____

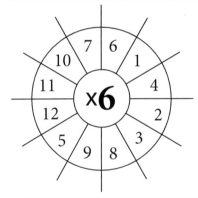 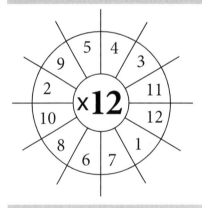 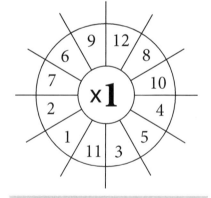

Correct:_____ Time:_____ # Correct:_____ Time:_____ # Correct:_____ Time:_____

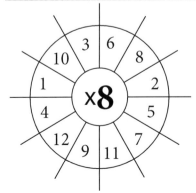  

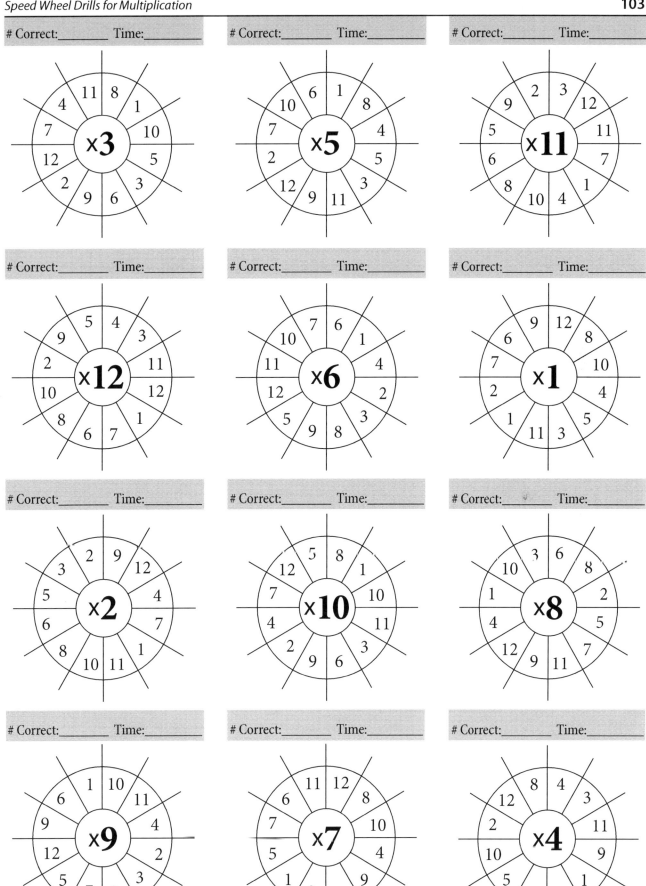

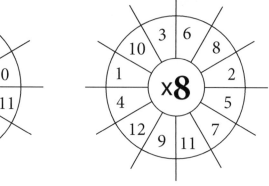
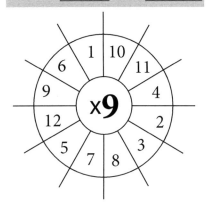

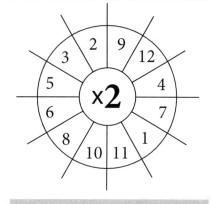
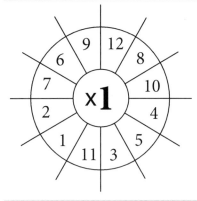
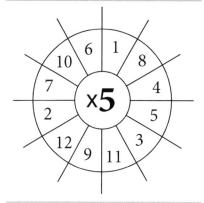

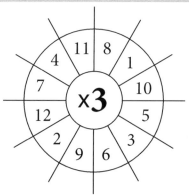

Speed Wheel Drills for Multiplication 105

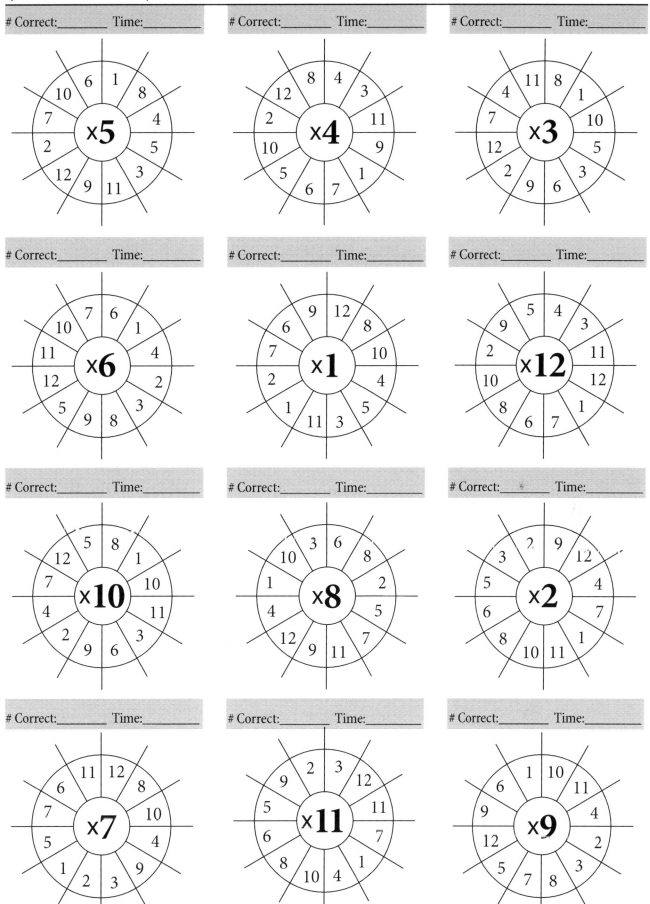

It is illegal to photocopy this page. Copyright © 2021, Richard W. Fisher

Correct:_____ Time:_____   # Correct:_____ Time:_____   # Correct:_____ Time:_____

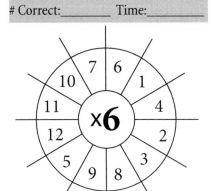 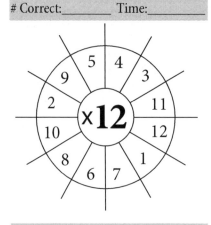 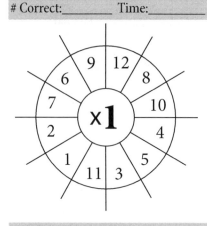

Correct:_____ Time:_____   # Correct:_____ Time:_____   # Correct:_____ Time:_____

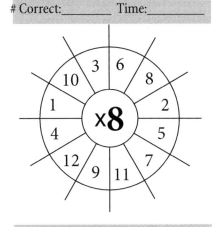 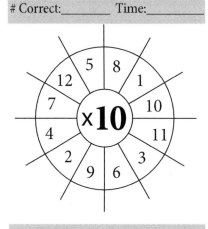 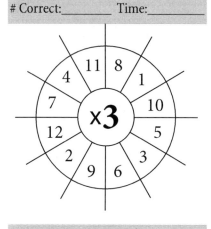

Correct:_____ Time:_____   # Correct:_____ Time:_____   # Correct:_____ Time:_____

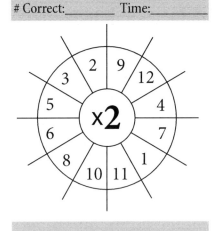 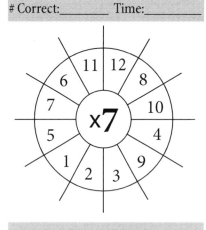 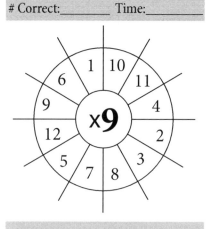

Correct:_____ Time:_____   # Correct:_____ Time:_____   # Correct:_____ Time:_____

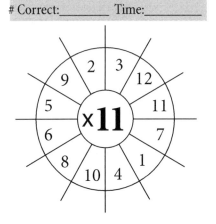 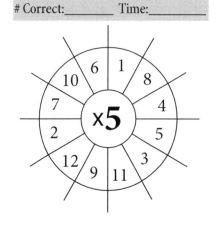 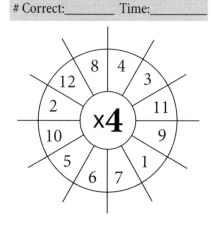

Speed Wheel Drills for Multiplication   107

# Correct:_____ Time:_____	# Correct:_____ Time:_____	# Correct:_____ Time:_____
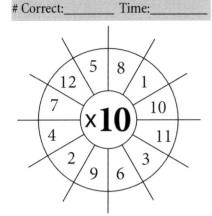		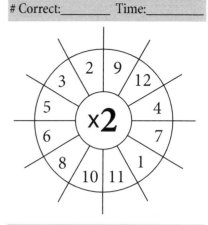

# Correct:_____ Time:_____	# Correct:_____ Time:_____	# Correct:_____ Time:_____
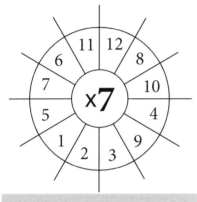		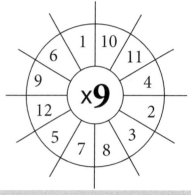

# Correct:_____ Time:_____	# Correct:_____ Time:_____	# Correct:_____ Time:_____
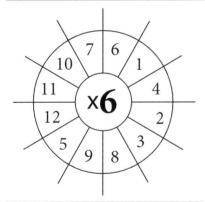		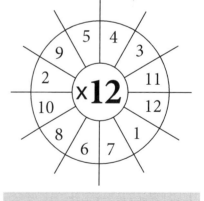

# Correct:_____ Time:_____	# Correct:_____ Time:_____	# Correct:_____ Time:_____
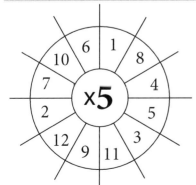		

It is illegal to photocopy this page. Copyright © 2021, Richard W. Fisher

Speed Wheel Drills for Multiplication

Correct:_____ Time:_____

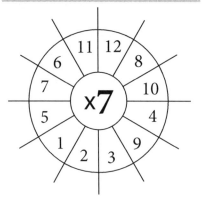

Correct:_____ Time:_____

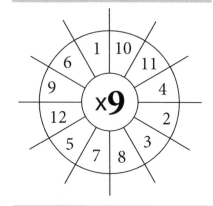

Correct:_____ Time:_____

Correct:_____ Time:_____

Correct:_____ Time:_____

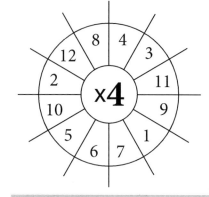

Correct:_____ Time:_____

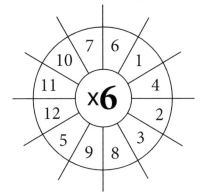

Correct:_____ Time:_____

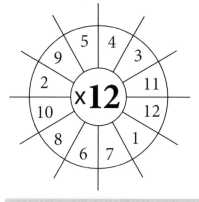

Correct:_____ Time:_____

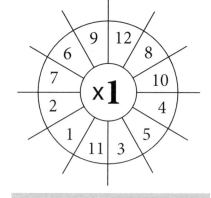

Correct:_____ Time:_____

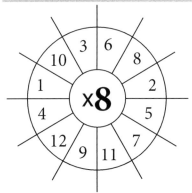

Correct:_____ Time:_____

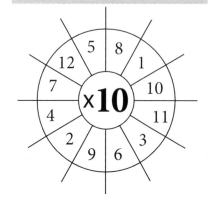

Correct:_____ Time:_____

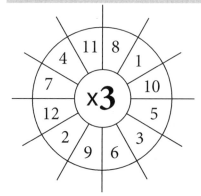

It is illegal to photocopy this page. Copyright © 2021, Richard W. Fisher

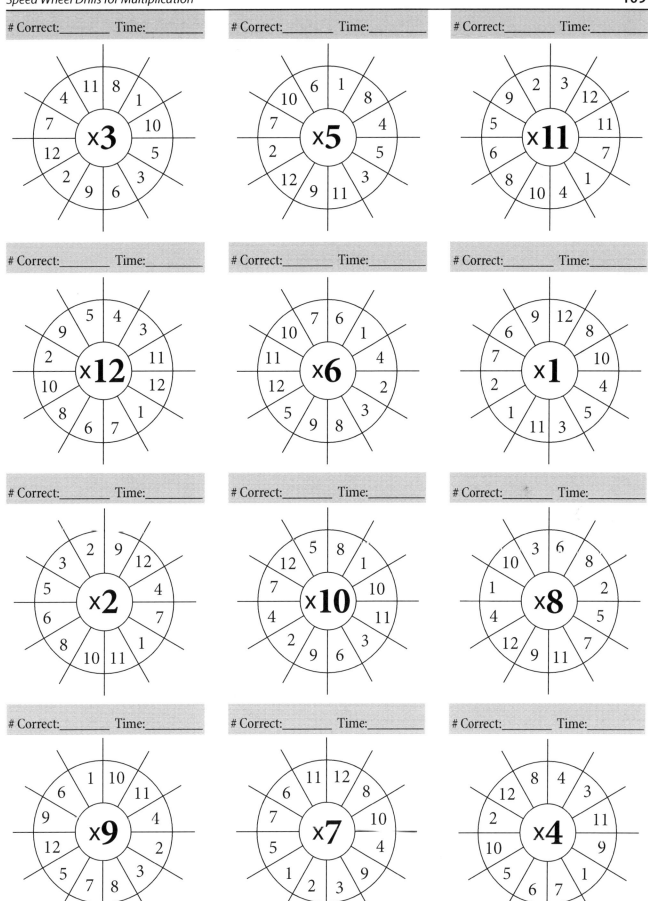

Correct:_____ Time:_____

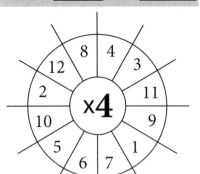

Correct:_____ Time:_____

Correct:_____ Time:_____

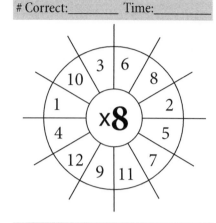

Correct:_____ Time:_____

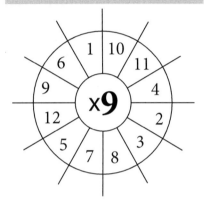

Correct:_____ Time:_____

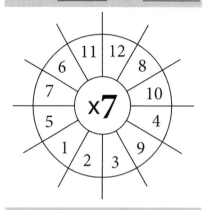

Correct:_____ Time:_____

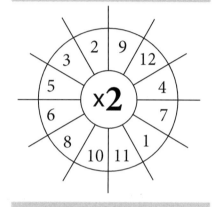

Correct:_____ Time:_____

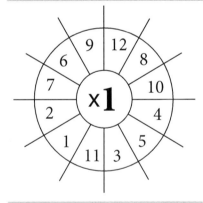

Correct:_____ Time:_____

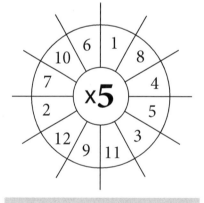

Correct:_____ Time:_____

Correct:_____ Time:_____

Correct:_____ Time:_____

Correct:_____ Time:_____

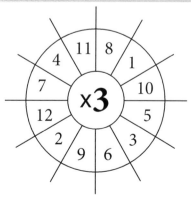

Speed Wheel Drills for Multiplication 111

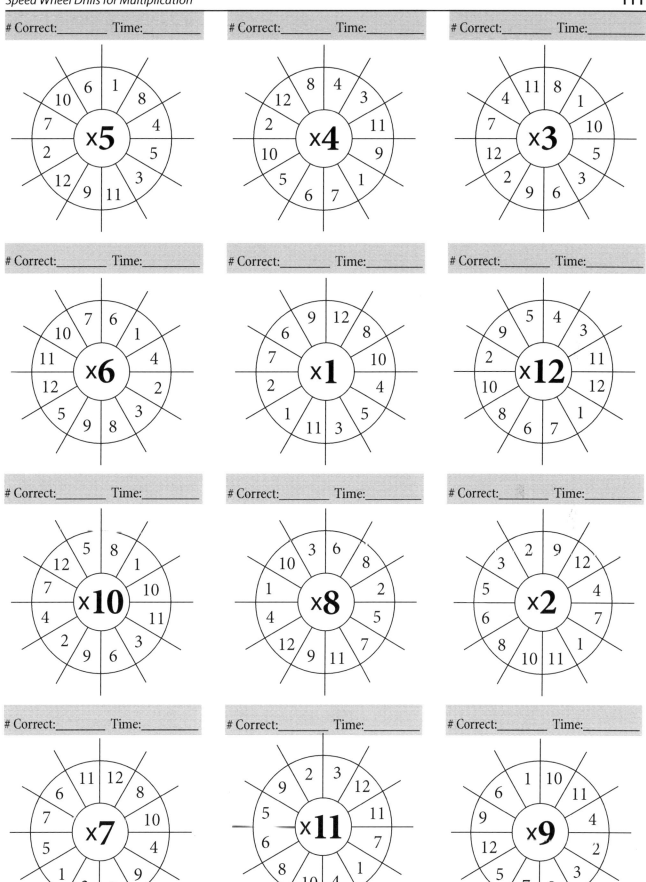

Correct:_____ Time:_____ # Correct:_____ Time:_____ # Correct:_____ Time:_____

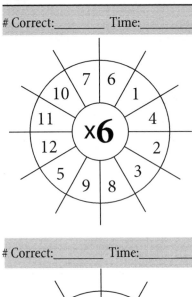

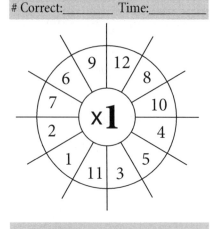

Correct:_____ Time:_____ # Correct:_____ Time:_____ # Correct:_____ Time:_____

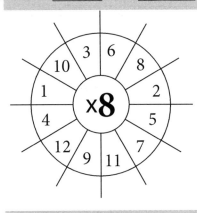

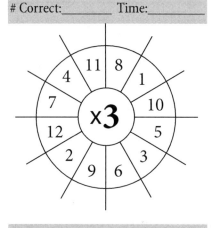

Correct:_____ Time:_____ # Correct:_____ Time:_____ # Correct:_____ Time:_____

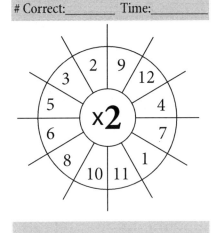
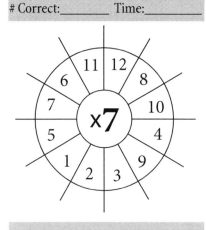
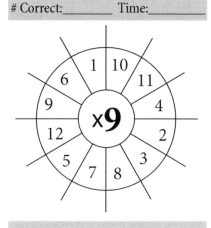

Correct:_____ Time:_____ # Correct:_____ Time:_____ # Correct:_____ Time:_____

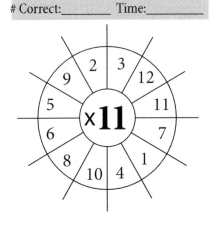
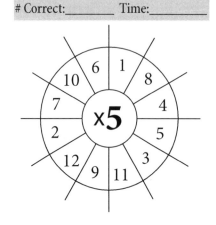
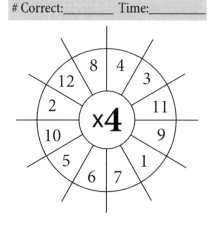

Speed Wheel Drills for Multiplication

Correct:_____ Time:_____

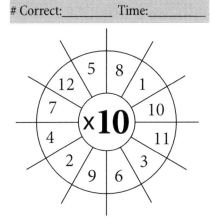

Correct:_____ Time:_____

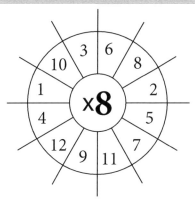

Correct:_____ Time:_____

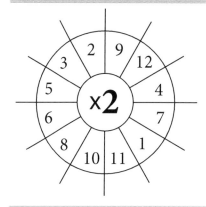

Correct:_____ Time:_____

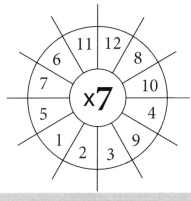

Correct:_____ Time:_____

Correct:_____ Time:_____

Correct:_____ Time:_____

Correct:_____ Time:_____

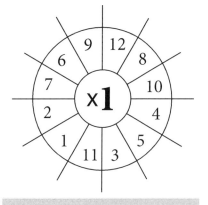

Correct:_____ Time:_____

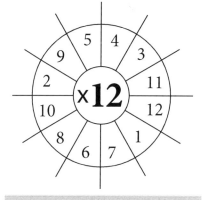

Correct:_____ Time:_____

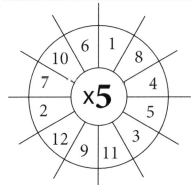

Correct:_____ Time:_____

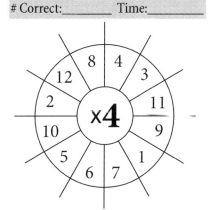

Correct:_____ Time:_____

It is illegal to photocopy this page. Copyright © 2021, Richard W. Fisher

Correct:_____ Time:_____
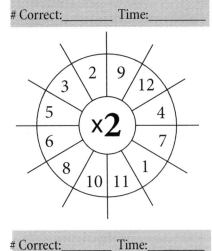

Correct:_____ Time:_____
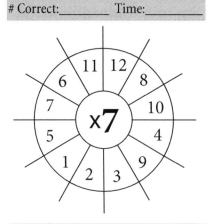

Correct:_____ Time:_____
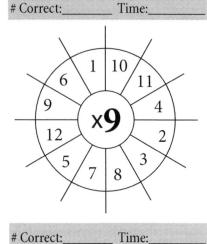

Correct:_____ Time:_____

Correct:_____ Time:_____
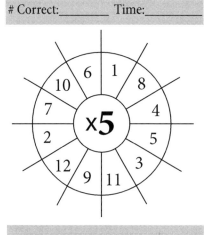

Correct:_____ Time:_____
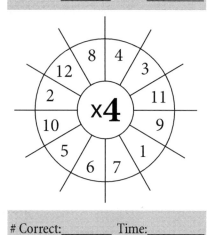

Correct:_____ Time:_____
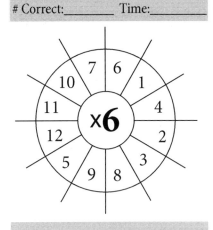

Correct:_____ Time:_____
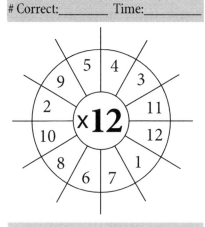

Correct:_____ Time:_____
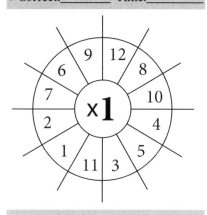

Correct:_____ Time:_____
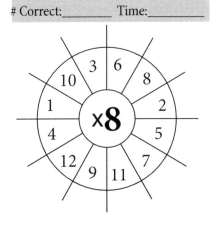

Correct:_____ Time:_____
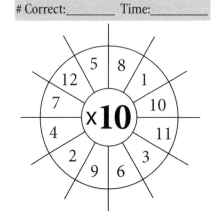

Correct:_____ Time:_____
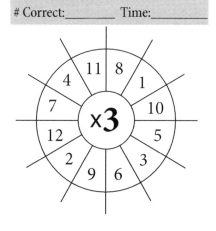

Speed Wheel Drills for Multiplication 115

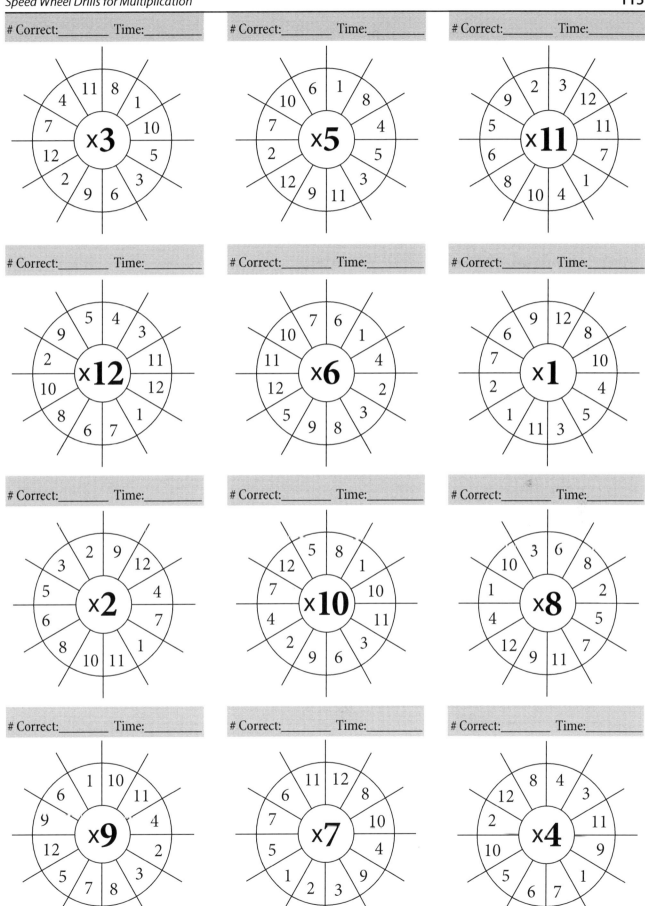

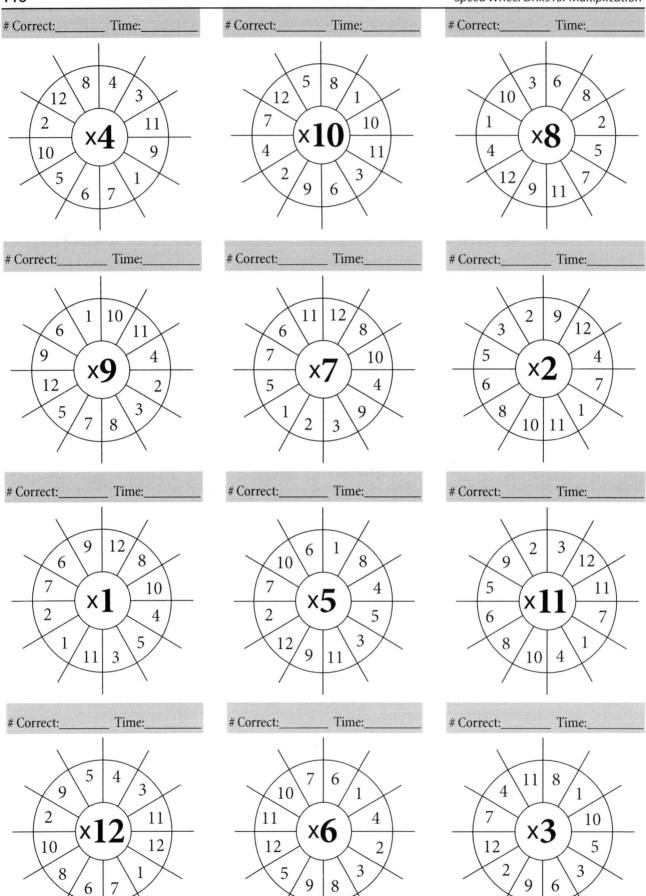

Speed Wheel Drills for Multiplication
117

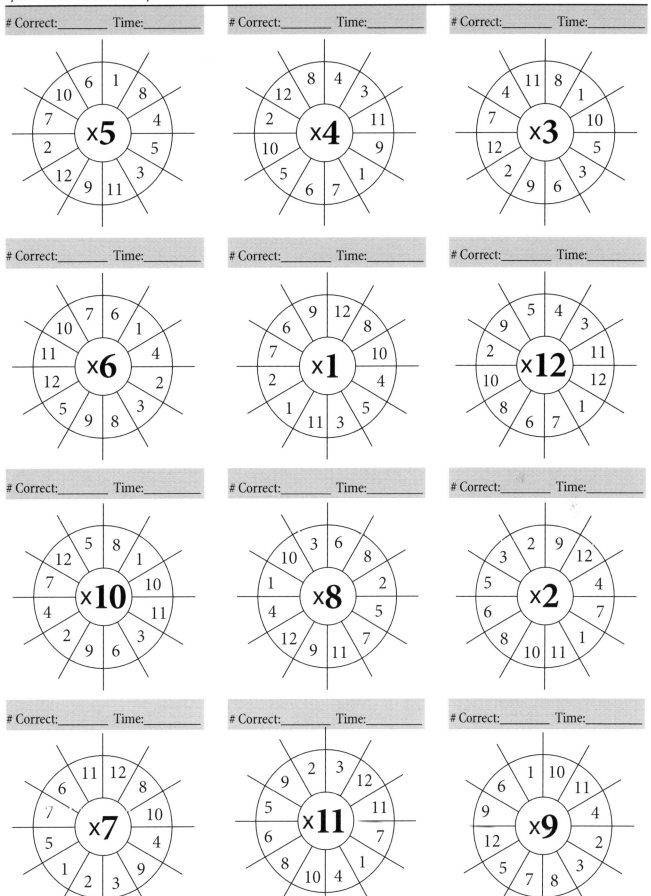

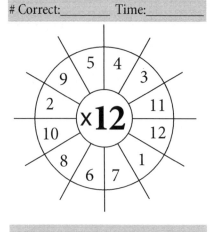
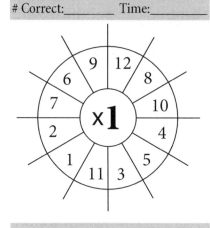
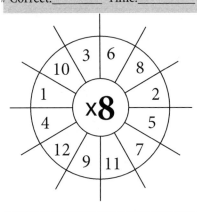
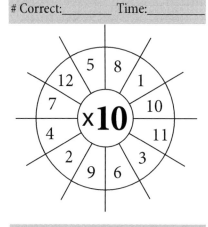
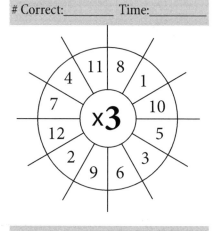
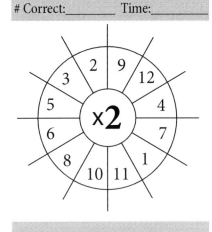

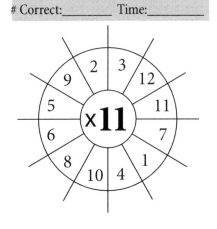
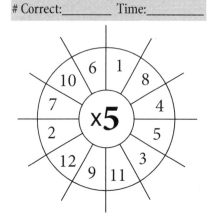
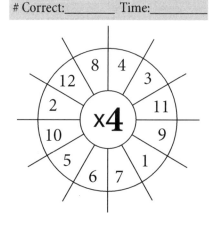

Speed Wheel Drills for Multiplication 119

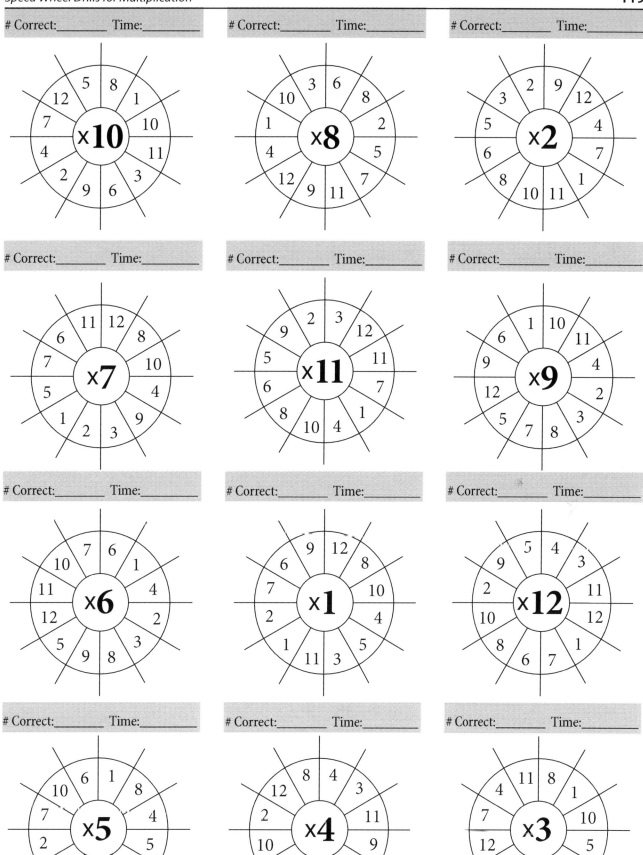

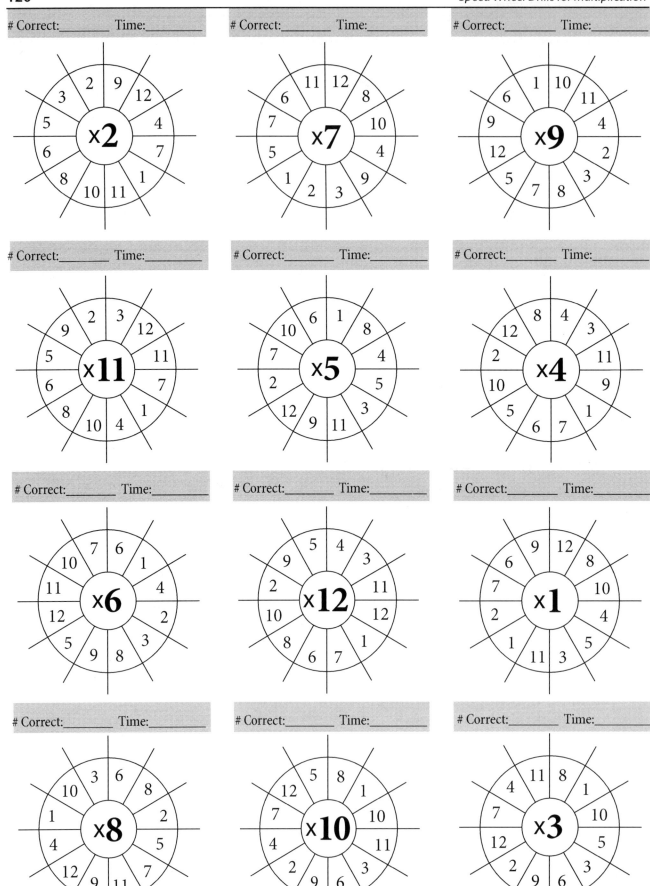

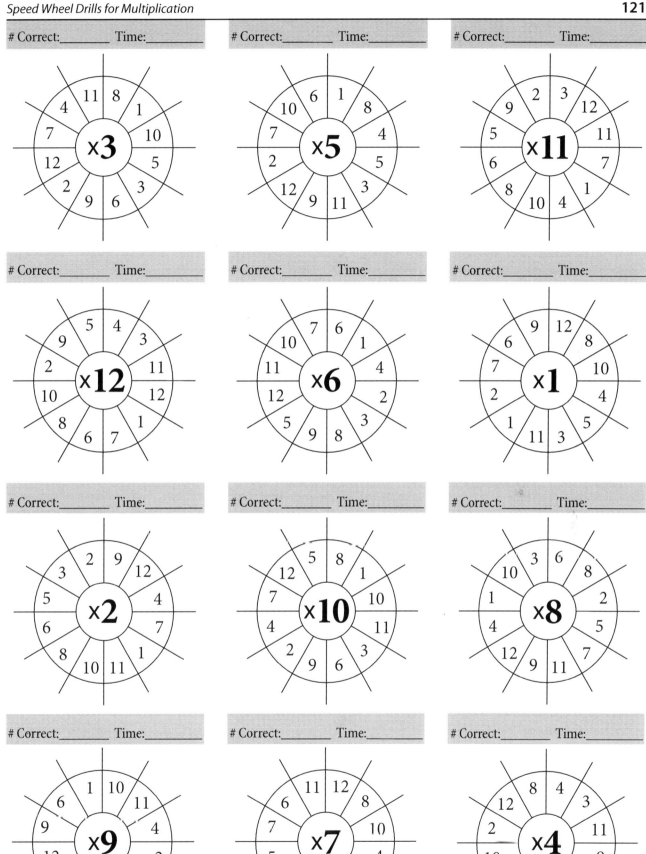

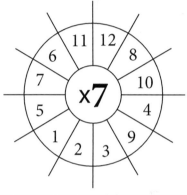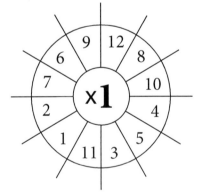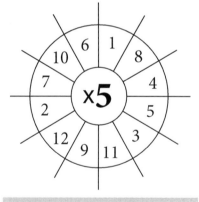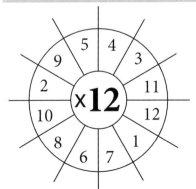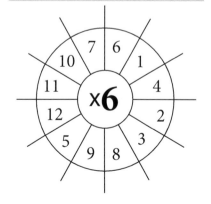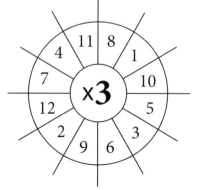

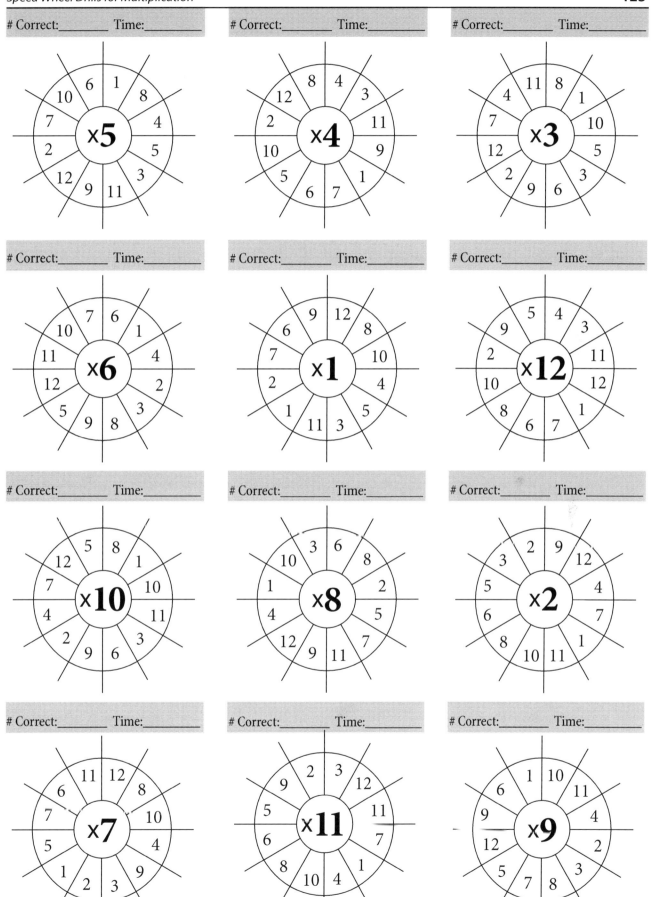

Speed Wheel Drills for Multiplication

Speed Wheel Drills for Multiplication   125

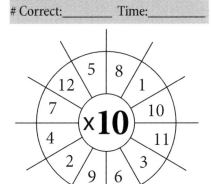
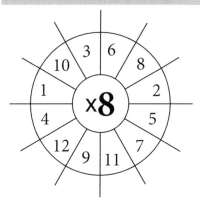
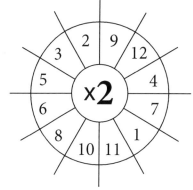

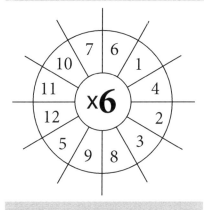
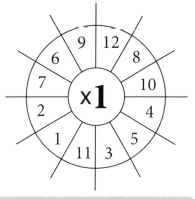

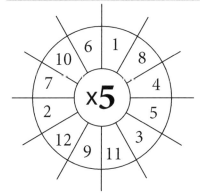

126 Speed Wheel Drills for Multiplication

Correct:_____ Time:_____
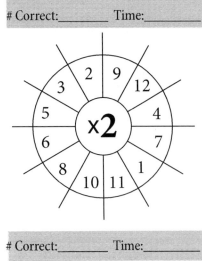

Correct:_____ Time:_____
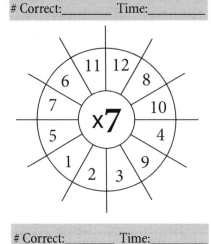

Correct:_____ Time:_____

Correct:_____ Time:_____
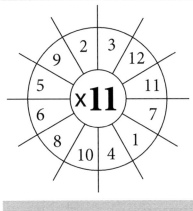

Correct:_____ Time:_____

Correct:_____ Time:_____

Correct:_____ Time:_____
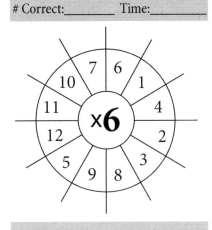

Correct:_____ Time:_____

Correct:_____ Time:_____
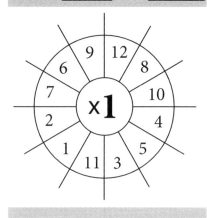

Correct:_____ Time:_____
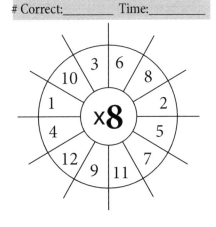

Correct:_____ Time:_____
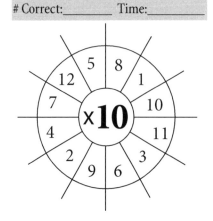

Correct:_____ Time:_____
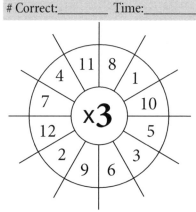

It is illegal to photocopy this page. Copyright © 2021, Richard W. Fisher

Resource Center

Math Glossary 129

Important Symbols 135

Mulipication Table 136

Commonly Used Prime Numbers.............. 136

Squares and Square Roots 137

Fraction/Decimal Equivalents 138

Speed Wheel Drill Solutions (Multiplication).... 139

A

absolute value   The distance of a number from 0 on the number line. The absolute value is always positive.

acute angle   An angle with a measure of less than 90 degrees.

adjacent   Next to.

algebraic expression   A mathematical expression that contains at least one variable.

angle   Any two rays that share an endpoint will form an angle.

associative properties   For any a, b, c:
addition: $(a + b) + c = a + (b + c)$
multiplication: $(ab)c = a(bc)$

B

base   The number being multiplied. In an expression such as 4^2, 4 is the base.

C

coefficient   A number that multiplies the variable. In the term 7x, 7 is the coefficient of x.

commutative properties   For any a, b:
addition: $a + b = b + a$
multiplication: $ab = ba$

complementary angles   Two angles that have measures whose sum is 90 degrees.

congruent   Two figures having exactly the same size and shape.

coordinate plane   The plane which contains the x- and y-axes. It is divided into 4 quadrants. Also called coordinate system and coordinate grid.

coordinates   An ordered pair of numbers that identify a point on a coordinate plane.

D

data   Information that is organized for analysis.

degree   A unit that is used in measuring angles.

denominator   The bottom number of a fraction that tells the number of equal parts into which a whole is divided.

disjoint sets   Sets that have no members in common. {1,2,3} and {4,5,6} are disjoint sets.

distributive property   For real numbers a, b, and c: a(b + c) = ab + ac.

E

element of a set   Member of a set.

empty set   The set that has no members. Also called the null set and written Ø or { }.

equation   A mathematical sentence that contains an equal sign (=) and states that one expression is equal to another expression.

equivalent   Having the same value.

exponent   A number that indicates the number of times a given base is used as a factor. In the expression n^2, 2 is the exponent.

expression   Variables, numbers, and symbols that show a mathematical relationship.

extremes of a proportion   In the proportion $\frac{a}{b} = \frac{c}{d}$, a and d are the extremes.

F

factor   An integer that divides evenly into another.

finite   Something that is countable.

formula   A general mathematical statement or rule. Used often in algebra and geometry.

function   A set of ordered pairs that pairs each x-value with one and only one y-value. (0,2), (-1,6), (4,-2), (-3,4) is a function.

G

graph   To show points named by numbers or ordered pairs on a number line or coordinate plane. Also, a drawing to show the relationship between sets of data.

greatest common factor   The largest common factor of two or more numbers. Also written GCF. The greatest common factor of 15 and 25 is 5.

grouping symbols   Symbols that indicate the order in which mathematical operations should take place. Examples include parentheses ( ), brackets [ ], braces { }, and fraction bars — .

H

hypotenuse   The side opposite the right angle in a right triangle.

I

identity properties of addition and multiplication For any real number a:
addition: a + 0 = 0 + a = a
multiplication: 1 × a = a × 1 = a

inequality A mathematical sentence that states one expression is greater than or less than another. Inequality symbols are read as follows: < less than
≤ less than or equal to
> greater than
≥ greater than or equal to

infinite Having no boundaries or limits. Uncountable.

integers Numbers in a set. ...-3, -2, -1, 0, 1, 2, 3...

intersection of sets If A and B are sets, then A intersection B is the set whose members are included in both sets A and B, and is written A ∩ B. If set A = {1,2,3,4} and set B = {1,3,5}, then A ∩ B = {1,3}

inverse properties of addition and multiplication For any number a:
addition: a + -a = 0
multiplication: a × 1/a = 1 (a ≠ 0)

inverse operations Operations that "undo" each other. Addition and subtraction are inverse operations, and multiplication and division are inverse operations.

L

least common multiple The least common multiple of two or more whole numbers is the smallest whole number, other than zero, that they all divide into evenly. Also written as LCM. The least common multiple of 12 and 8 is 24.

linear equation An equation whose graph is a straight line.

M

mean In statistics, the sum of a set of numbers divided by the number of elements in the set. Sometimes referred to as average.

means of a proportion In the proportion $\frac{a}{b} = \frac{c}{d}$, b and c are the means.

median In statistics, the middle number of a set of numbers when the numbers are arranged in order of least to greatest. If there are two middle numbers, find their mean.

mode In statistics, the number that appears most frequently. Sometimes there is no mode. There may also be more than one mode.

multiple The product of a whole number and another whole number.

N

natural numbers   Numbers in the set 1, 2, 3, 4,... Also called counting numbers.

negative numbers   Numbers that are less than zero.

null set   The set that has no members. Also called the empty set and written Ø or { }.

number line   A line that represents numbers as points.

numerator   The top part of a fraction.

O

obtuse angle   An angle whose measure is greater than 90° and less than 180°.

opposites   Numbers that are the same distance from zero, but are on opposite sides of zero on a number line. 4 and -4 are opposites.

order of operations   The order of steps to be used when simplifying expressions.
   1. Evaluate within grouping symbols.
   2. Eliminate all exponents.
   3. Multiply and divide in order from left to right.
   4. Add and subtract in order from left to right.

ordered pair   A pair of numbers (x,y) that represent a point on the coordinate plane. The first number is the x-coordinate and the second number is the y-coordinate.

origin   The point where the x-axis and the y-axis intersect in a coordinate plane. Written as (0,0).

outcome   One of the possible events in a probability situation.

P

parallel lines   Lines in a plane that do not intersect. They stay the same distance apart.

percent   Hundredths or per hundred. Written %.

perimeter   The distance around a figure.

perpendicular lines   Lines in the same plane that intersect at a right (90°) angle.

pi   The ratio of the circumference of a circle to its diameter. Written π. The approximate value for π is 3.14 as a decimal and $\frac{22}{7}$ as a fraction.

plane   A flat surface that extends infinitely in all directions.

point   An exact position in space. Points also represent numbers on a number line or coordinate plane.

positive number   Any number that is greater than 0.

power   An exponent.

prime number   A whole number greater than 1 whose only factors are 1 and itself.

probability   What chance, or how likely it is for an event to occur. It is the ratio of the ways a certain outcome can occur and the number of possible outcomes.

proportion   An equation that states that two ratios are equal. $\frac{4}{8} = \frac{2}{4}$ is a proportion.

Pythagorean theorem   In a right triangle, if c is the hypotenuse, and a and b are the other two legs, then $a^2 + b^2 = c^2$.

Q

quadrant   One of the four regions into which the x-axis and y-axis divide a coordinate plane.

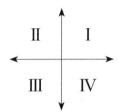

R

range   The difference between the greatest number and the least number in a set of numbers.

ratio   A comparison of two numbers using division. Written a:b, a to b, and a/b.

reciprocals   Two numbers whose product is 1. $\frac{2}{3}$ and $\frac{3}{2}$ are reciprocals because $\frac{2}{3} \times \frac{3}{2} = 1$.

reduce   To express a fraction in its lowest terms.

relation   Any set of ordered pairs.

right angle   An angle that has a measure of 90°.

rise   The change in y going from one point to another on a coordinate plane. The vertical change.

run   The change in x going from one point to another on a coordinate plane. The horizontal change.

S

scientific notation   A number written as the product of a numbers between 1 and 10 and a power of ten. In scientific notation, $7,000 = 7 \times 10^3$.

set   A well-defined collection of objects.

slope   Refers to the slant of a line. It is the ratio of rise to run.

solution   A number that can be substituted for a variable to make an equation true.

square root   Written $\sqrt{}$. The $\sqrt{36} = 6$ because $6 \times 6 = 36$.

statistics   Involves data that is gathered about people or things and is used for analysis.

subset   If all the members of set A are members of set B, then set A is a subset of set B. Written $A \subset B$. If set A = {1,2,3} and set B = {0,1,2,3,5,8}, set A is a subset of set B because all of the members of a set A are also members of set B.

U

union of sets   If A and B are sets, the union of set A and set B is the set whose members are included in set A, or set B, or both set A and set B. A union B is written $A \cup B$. If set = {1,2,3,4} and set B = {1,3,5,7}, then $A \cup B$ = {1,2,3,4,5,7}.

universal set   The set which contains all the other sets which are under consideration.

V

variable   A letter that represents a number.

Venn diagram   A type of diagram that shows how certain sets are related.

vertex   The point at which two lines, line segments, or rays meet to form an angle.

W

whole number   Any number in the set 0, 1, 2, 3, 4...

X

x-axis   The horizontal axis on a coordinate plane.

x-coordinate   The first number in an ordered pair. Also called the abscissa.

Y

y-axis   The vertical axis on a coordinate plane.

y-coordinate   The second number in an ordered pair. Also called the ordinate.

Important Symbols

$<$	less than	π	pi
$\leq$	less than or equal to	$\{\ \}$	set
$>$	greater than	$|\ |$	absolute value
$\geq$	greater than or equal to	$.\overline{n}$	repeating decimal symbol
$=$	equal to	$1/a$	the reciprocal of a number
$\neq$	not equal to	$\%$	percent
$\cong$	congruent to	(x,y)	ordered pair
$(\ )$	parenthesis	$\perp$	perpendicular
$[\ ]$	brackets	$||$	parallel to
$\{\ \}$	braces	$\angle$	angle
$...$	and so on	$\in$	element of
$\cdot$ or $\times$	multiply	$\notin$	not an element of
∞	infinity	$\cap$	intersection
a^n	the nth power of a number	$\cup$	union
$\sqrt{\ }$	square root	$\subset$	subset of
$\emptyset, \{\ \}$	the empty set or null set	$\not\subset$	not a subset of
$\therefore$	therefore	$\triangle$	triangle
$°$	degree		

Multiplication Table

x	2	3	4	5	6	7	8	9	10	11	12
2	4	6	8	10	12	14	16	18	20	22	24
3	6	9	12	15	18	21	24	27	30	33	36
4	8	12	16	20	24	28	32	36	40	44	48
5	10	15	20	25	30	35	40	45	50	55	60
6	12	18	24	30	36	42	48	54	60	66	72
7	14	21	28	35	42	49	56	63	70	77	84
8	16	24	32	40	48	56	64	72	80	88	96
9	18	27	36	45	54	63	72	81	90	99	108
10	20	30	40	50	60	70	80	90	100	110	120
11	22	33	44	55	66	77	88	99	110	121	132
12	24	36	48	60	72	84	96	108	120	132	144

Commonly Used Prime Numbers

2	3	5	7	11	13	17	19	23	29
31	37	41	43	47	53	59	61	67	71
73	79	83	89	97	101	103	107	109	113
127	131	137	139	149	151	157	163	167	173
179	181	191	193	197	199	211	223	227	229
233	239	241	251	257	263	269	271	277	281
283	293	307	311	313	317	331	337	347	349
353	359	367	373	379	383	389	397	401	409
419	421	431	433	439	443	449	457	461	463
467	479	487	491	499	503	509	521	523	541
547	557	563	569	571	577	587	593	599	601
607	613	617	619	631	641	643	647	653	659
661	673	677	683	691	701	709	719	727	733
739	743	751	757	761	769	773	787	797	809
811	821	823	827	829	839	853	857	859	863
877	881	883	887	907	911	919	929	937	941
947	953	967	971	977	983	991	997	1009	1013

Squares and Square Roots

No.	Square	Square Root	No.	Square	Square Root	No.	Square	Square Root
1	1	1.000	51	2,601	7.141	101	10201	10.050
2	4	1.414	52	2,704	7.211	102	10,404	10.100
3	9	1.732	53	2,809	7.280	103	10,609	10.149
4	16	2.000	54	2,916	7.348	104	10,816	10.198
5	25	2.236	55	3,025	7.416	105	11,025	10.247
6	36	2.449	56	3,136	7.483	106	11,236	10.296
7	49	2.646	57	3,249	7.550	107	11,449	10.344
8	64	2.828	58	3,364	7.616	108	11,664	10.392
9	81	3.000	59	3,481	7.681	109	11,881	10.440
10	100	3.162	60	3,600	7.746	110	12,100	10.488
11	121	3.317	61	3,721	7.810	111	12,321	10.536
12	144	3.464	62	3,844	7.874	112	12,544	10.583
13	169	3.606	63	3,969	7.937	113	12,769	10.630
14	196	3.742	64	4,096	8.000	114	12,996	10.677
15	225	3.873	65	4,225	8.062	115	13,225	10.724
16	256	4.000	66	4,356	8.124	116	13,456	10.770
17	289	4.123	67	4,489	8.185	117	13,689	10.817
18	324	4.243	68	4,624	8.246	118	13,924	10.863
19	361	4.359	69	4,761	8.307	119	14,161	10.909
20	400	4.472	70	4,900	8.367	120	14,400	10.954
21	441	4.583	71	5,041	8.426	121	14,641	11.000
22	484	4.690	72	5,184	8.485	122	14,884	11.045
23	529	4.796	73	5,329	8.544	123	15,129	11.091
24	576	4.899	74	5,476	8.602	124	15,376	11.136
25	625	5.000	75	5,625	8.660	125	15,625	11.180
26	676	5.099	76	5,776	8.718	126	15,876	11.225
27	729	5.196	77	5,929	8.775	127	16,129	11.269
28	784	5.292	78	6,084	8.832	128	16,384	11.314
29	841	5.385	79	6,241	8.888	129	16,641	11.358
30	900	5.477	80	6,400	8.944	130	16,900	11.402
31	961	5.568	81	6,561	9.000	131	17,161	11.446
32	1,024	5.657	82	6,724	9.055	132	17,424	11.489
33	1,089	5.745	83	6,889	9.110	133	17,689	11.533
34	1,156	5.831	84	7,056	9.165	134	17,956	11.576
35	1,225	5.916	85	7,225	9.220	135	18,225	11.619
36	1,296	6.000	86	7,396	9.274	136	18,496	11.662
37	1,369	6.083	87	7,569	9.327	137	18,769	11.705
38	1,444	6.164	88	7,744	9.381	138	19,044	11.747
39	1,521	6.245	89	7,921	9.434	139	19,321	11.790
40	1,600	6.325	90	8,100	9.487	140	19,600	11.832
41	1,681	6.403	91	8,281	9.539	141	19,881	11.874
42	1,764	6.481	92	8,464	9.592	142	20,164	11.916
43	1,849	6.557	93	8,649	9.644	143	20,449	11.958
44	1,936	6.633	94	8,836	9.695	144	20,736	12.000
45	2,025	6.708	95	9,025	9.747	145	21,025	12.042
46	2,116	6.782	96	9,216	9.798	146	21,316	12.083
47	2,209	6.856	97	9,409	9.849	147	21,609	12.124
48	2,304	6.928	98	9,604	9.899	148	21,904	12.166
49	2,401	7.000	99	9,801	9.950	149	22,201	12.207
50	2,500	7.071	100	10,000	10.000	150	22,500	12.247

Fraction/Decimal Equivalents

Fraction	Decimal	Fraction	Decimal
$\frac{1}{2}$	0.5	$\frac{5}{10}$	0.5
$\frac{1}{3}$	$0.\overline{3}$	$\frac{6}{10}$	0.6
$\frac{2}{3}$	$0.\overline{6}$	$\frac{7}{10}$	0.7
$\frac{1}{4}$	0.25	$\frac{8}{10}$	0.8
$\frac{2}{4}$	0.5	$\frac{9}{10}$	0.9
$\frac{3}{4}$	0.75	$\frac{1}{16}$	0.0625
$\frac{1}{5}$	0.2	$\frac{2}{16}$	0.125
$\frac{2}{5}$	0.4	$\frac{3}{16}$	0.1875
$\frac{3}{5}$	0.6	$\frac{4}{16}$	0.25
$\frac{4}{5}$	0.8	$\frac{5}{16}$	0.3125
$\frac{1}{8}$	0.125	$\frac{6}{16}$	0.375
$\frac{2}{8}$	0.25	$\frac{7}{16}$	0.4375
$\frac{3}{8}$	0.375	$\frac{8}{16}$	0.5
$\frac{4}{8}$	0.5	$\frac{9}{16}$	0.5625
$\frac{5}{8}$	0.625	$\frac{10}{16}$	0.625
$\frac{6}{8}$	0.75	$\frac{11}{16}$	0.6875
$\frac{7}{8}$	0.875	$\frac{12}{16}$	0.75
$\frac{1}{10}$	0.1	$\frac{13}{16}$	0.8125
$\frac{2}{10}$	0.2	$\frac{14}{16}$	0.875
$\frac{3}{10}$	0.3	$\frac{15}{16}$	0.9375
$\frac{4}{10}$	0.4		

Speed Wheel Drill Solutions (Multiplication)

x1	x2	x3	x4	x5	x6
1 x 1 = 1	2 x 1 = 2	3 x 1 = 3	4 x 1 = 4	5 x 1 = 5	6 x 1 = 6
1 x 2 = 2	2 x 2 = 4	3 x 2 = 6	4 x 2 = 8	5 x 2 = 10	6 x 2 = 12
1 x 3 = 3	2 x 3 = 6	3 x 3 = 9	4 x 3 = 12	5 x 3 = 15	6 x 3 = 18
1 x 4 = 4	2 x 4 = 8	3 x 4 = 12	4 x 4 = 16	5 x 4 = 20	6 x 4 = 24
1 x 5 = 5	2 x 5 = 10	3 x 5 = 15	4 x 5 = 20	5 x 5 = 25	6 x 5 = 30
1 x 6 = 6	2 x 6 = 12	3 x 6 = 18	4 x 6 = 24	5 x 6 = 30	6 x 6 = 36
1 x 7 = 7	2 x 7 = 14	3 x 7 = 21	4 x 7 = 28	5 x 7 = 35	6 x 7 = 42
1 x 8 = 8	2 x 8 = 16	3 x 8 = 24	4 x 8 = 32	5 x 8 = 40	6 x 8 = 48
1 x 9 = 9	2 x 9 = 18	3 x 9 = 27	4 x 9 = 36	5 x 9 = 45	6 x 9 = 54
1 x 10 = 10	2 x 10 = 20	3 x 10 = 30	4 x 10 = 40	5 x 10 = 50	6 x 10 = 60
1 x 11 = 11	2 x 11 = 22	3 x 11 = 33	4 x 11 = 44	5 x 11 = 55	6 x 11 = 66
1 x 12 = 12	2 x 12 = 24	3 x 12 = 36	4 x 12 = 48	5 x 12 = 60	6 x 12 = 72

x7	x8	x9	x10	x11	x12
7 x 1 = 7	8 x 1 = 8	9 x 1 = 9	10 x 1 = 10	11 x 1 = 11	12 x 1 = 12
7 x 2 = 14	8 x 2 = 16	9 x 2 = 18	10 x 2 = 20	11 x 2 = 22	12 x 2 = 24
7 x 3 = 21	8 x 3 = 24	9 x 3 = 27	10 x 3 = 30	11 x 3 = 33	12 x 3 = 36
7 x 4 = 28	8 x 4 = 32	9 x 4 = 36	10 x 4 = 40	11 x 4 = 44	12 x 4 = 48
7 x 5 = 35	8 x 5 = 40	9 x 5 = 45	10 x 5 = 50	11 x 5 = 55	12 x 5 = 60
7 x 6 = 42	8 x 6 = 48	9 x 6 = 54	10 x 6 = 60	11 x 6 = 66	12 x 6 = 72
7 x 7 = 49	8 x 7 = 56	9 x 7 = 63	10 x 7 = 70	11 x 7 = 77	12 x 7 = 84
7 x 8 = 56	8 x 8 = 64	9 x 8 = 72	10 x 8 = 80	11 x 8 = 88	12 x 8 = 96
7 x 9 = 63	8 x 9 = 72	9 x 9 = 81	10 x 9 = 90	11 x 9 = 99	12 x 9 = 108
7 x 10 = 70	8 x 10 = 80	9 x 10 = 90	10 x 10 = 100	11 x 10 = 110	12 x 10 = 120
7 x 11 = 77	8 x 11 = 88	9 x 11 = 99	10 x 11 = 110	11 x 11 = 121	12 x 11 = 132
7 x 12 = 84	8 x 12 = 96	9 x 12 = 108	10 x 12 = 120	11 x 12 = 132	12 x 12 = 144

It is illegal to photocopy this page. Copyright © 2021, Richard W. Fisher

Made in the USA
Monee, IL
08 September 2022

13504421R00079